I0473471

www.ingramcontent.com/pod-product-compliance
Lightning Source LLC
Chambersburg PA
CBHW070437180526
45158CB00019B/1471

A8 The Heroes

The heroes of this story, to which I offer my gratitude, are listed below

It is not necessary to identify the invaluable contributions made by each of these contributors, they are all widely known and available in almost every scientific publication in circulation today.

Nicolaus Copernicus (Polish) 1473-1543
William Gilbert (English) 1544-1603
Tyco Brahe (Danish) 1546-1601
Galileo Galilei (Italian) 1564-1642
Johannes Kepler (German) 1571-1630
Christiaan Huygens (Dutch) 1629-1695
Isaac Newton (English) 1642-1727
Edmund Halley (English) 1656-1741
Charles-Augustin de Coulomb (French) 1736-1806
Hans Christian Ørsted (Danish) 1777-1851
Michael Faraday (English) 1791-1867
Josef Stefan (Austria) 1815-1863
James Clerk Maxwell (Scottish) 1831-1879
William Crookes (English) 1832-1919
Ludwig Boltzmann (Austria) 1844-1906
Hendrik Lorentz (Dutch) 1853-1928
Jules Henri Poincaré (French) 1854-1912
Johannes Robert Rydberg (Swedish) 1854-1919
Max Karl Ernst Ludwig Planck (German) 1858-1947

The others that were instrumental in the completion of this book are:

My long-suffering wife (Brigitte) sub-editor and critic

My daughter (Eléonore), who initiated this project

Kenneth Pickering friend & editor, who first suggested that I write it

My thanks go out to all the above each of whom have provided a valuable piece of the puzzle without which the final solution would not have been possible, along with my sincere apologies to anybody I have unintentionally omitted.

A7 Hypotheses

Hypothesis 1: There is no such thing as mass.

Hypothesis 2: The earth's magnetic field reverses when a galactic comet passes sufficiently close to tip the earth on its axis.

Hypothesis 3: Electrons gain kinetic energy from electro-magnetic radiation, but they can only lose it via proton-electron pairing or impact.

Hypothesis 4: A lattice structure is mirrored in the atomic nucleic matrix.

Hypothesis 5: Only atoms of identical nucleic construction can generate lattice structures.

Hypothesis 6: An element's lattice structure also applies to its gaseous form and is responsible for partial pressure.

PHILOSOPHIÆ NATURALIS

PRINCIPIA MATHEMATICA

Revision IV – Vol III

By

Isaac Newton

And

Keith Dixon-Roche

PHILOSOPHIÆ NATURALIS

PRINCIPIA MATHEMATICA

Revision IV – Vol III

Published by CalQlata

info@CalQlata.com

First published November 2018
Second publication February 2019
Third publication June 2019
Fourth publication December 2019
Copyright © Keith Dixon-Roche 2018

Contents

Preface

I have always believed that if a mathematical law applies to one feature of nature it must apply to all of it: i.e. a law must by definition, be universal. I also feel that science took a wrong turning in the first quarter of the twentieth century owing to the dissemination of highly speculative theories that were accepted simply because of the prominence of their proposers. However, I was not sufficiently familiar with the subject to dispute it. After two and a half years of detailed study, that situation has changed and it appears to me that a hundred years may have been wasted in the search for impossible solutions. Isaac Newton's laws should have prevailed.

Newton apparently devised his theories to settle a bet, and like everything he tackled he took this work seriously. Despite having only Kepler's elliptical orbits and Galileo's laws of motion at his disposal, Newton managed to develop an all-encompassing theory that remains universally valid today. It was published in three revisions between 1687 and posthumously. He published only because of the persistence of one of his few friends: Edmund Halley. It is for this, rather than for his comet, that we owe Edmund Halley our deepest gratitude.

Newton's gravitational theory is complete and totally accurate. It covers all the bases. His model relies on a concept he called his 'constant of motion' to keep things moving. However, even he didn't realise that his theory also applies to circular orbits in which a satellite (e.g. an electron) provides its own kinetic energy, or that his gravitational constant (G) may be used to calculate the deflection of electro-magnetic radiation (light).

His laws of gravitation and motion together describe the behaviour of everything in the universe from atomic particles to the Big Bang, and they do so with absolute simplicity and accuracy, except for one small omission; he did not explain spin theory, without which it is difficult to explain all motion. However, it would have been very difficult for him to have developed this theory with the limited information and facilities available to him.

A couple of years ago, my daughter gave me a copy of Colin Pask's *Magnificent Principia* (Pask; 2013). After reading it, I was left with the suspicion that there were many unanswered questions about Newton's discoveries and I wondered how much had been done to continue Newton's work during the subsequent 300 years. Very little it seems.

I therefore set out on my quest to prove every aspect of Newton's theory of orbital motion, and see if I could determine the source of planetary spin. Having completed these objectives, I continued with core pressure, the earth's magnetic field, the definition of his gravitational constant 'G' and finally the atom, all using his theories.

After completing my model of the atom and having discovered how it really works, I was stunned by its simplicity and brilliance. Its existence must surely be due to providence, not chance. If there is one thing that could prove the existence of a being of supreme intelligence, and I am not referring to anybody's particular god; it is the atom. In the immortal words of a great contemporary philosopher; I was that *"girl sitting on her own in a small café in Rickmansworth"* (Adams; 1980), and I couldn't understand why none of this had been done before.

I am an engineer, not a scientist. Whilst I have always had an interest in science, I never had the opportunity to study it in detail. As a non-scientist, I have been able to tackle the subject free from the dogma that the scientific community has acquired since it displaced the religious community's hegemony over its own flawed natural laws.

Whilst my theories and models may not be perfect, everything in them can be supported with known scientific theories evolved well before the twentieth century.

I realise of course, that just as with Copernicus, Kepler, Galileo, Newton and Wegener before, none of these findings will be appreciated whilst the current scientific community exists. That august body is hardly likely to accept theories that disprove those for which they have been awarding themselves so many prestigious prizes. My hope is that maybe, one day, a new generation of free-thinking scientists will discover this work, correct, complete and advance it, and in so doing get science back on track.

Because it is now possible to define the Milky-Way's force-centre, I have given it the name 'Hades' for easier reference.

Keith Dixon-Roche 2018

1 Introduction

Science got itself into a bit of a mess during the 20th century owing to a couple of obscure theories, neither of which can be reconciled with concepts that we *know* work, but which stubbornly refuse to go away. Together these theories have inspired countless myths that simply multiply with the passing years. Nobody appears to be questioning them and nobody is able to verify them.

It has now become standard practice within the scientific community to justify any irreconcilable theory simply by claiming that *"the laws of physics do not apply"*.

So, I decided to have a go myself, by starting all over again; going back to basics (the year 1900).

Apart from Max Planck's assistance, I have managed to sort out this mess and compile a complete working theory for the universe using principles that were available well before 1900.

I have also managed to describe all the universal constants (including electrical) in the same basic units of energy (mass, length, time, charge & temperature; refer to Chapter 5)

Anything in this book that has not been fully resolved (and there isn't much) is referred to as *hypothesis*.

Whilst my hypotheses are perfectly robust, they remain as such because a couple of details need confirming/correcting. The contentious aspects mostly involve the nature of neutrons, but this has not been addressed here because it has nothing to do with Newton's laws of orbital motion.

Unresolved issues are highlighted in the text with the superscript [?] in which '?' will be replaced with a number that can be found in Chapter 7

1.1   What Went Wrong

Unfortunately, about a hundred years ago, a prominent scientist stated of his own theoretical model: *"if you aren't profoundly shocked by quantum physics, then you haven't understood it"*
Another did not appear to understand the basis for Henri Poincaré's formula E=mc² and actually declared to Georges Lemaître that his (Lemaître's) *"science was not very good"*

Such comments should be treated with extreme caution ...
... **if there is one thing certain about nature, it does not need to rely on complexity for an elegant solution**, and scientific laws *never* rely on statistics because statistics are subject to change; *laws are not.*
Statistics are akin to chaos theory: they are a means of guesswork used in situations where insufficient information is available to explain events accurately. They apply to the consequences of laws, not the laws themselves.

Quantum theory is inelegant, over complicated, reliant on statistics, cannot be reconciled with Newton's laws of orbital motion, cannot emit energy and remains unresolved after a hundred years. It is highly likely therefore, that it is nothing more than an obscure theoretical exercise.

Relativism can be disproved using Newton's gravitational constant and dark matter remains undiscovered. Poincaré's formula has nothing to do with kinetics. Classical atomic theory appears to be incorrect. Black-holes are wrongly said to be singularities that spin at the speed of light. Nobody has tried to determine the source of planetary (and therefore atomic) spin or core-pressure.

Isaac Newton pointed us in the right direction 300 years ago, but since the early 20th century the entire scientific community seems to have discounted the suitability of his theories for the evaluation of atoms (quantum theory) and galaxies (dark matter) simply because a couple of well-known scientists took this view. For example:

Relativism appears to have been partly based upon the supposition that E=mc² applies to kinematics, whereas it is a limiting case for potential energy based upon Newton's and Coulomb's laws and the creation of neutrons. Moreover, it incorrectly assumes that light possesses mass.

It is incorrectly currently believed that mass converts to energy with speed.

Dark matter in the form of sub-atomic particles was postulated because Newton's laws were said to predict a great deal more matter in the Milky Way than appears through observation. This has been easily disproved.

It was long ago assumed that we need sub-atomic particles (e.g. quarks, leptons, fermions, bosons, gluons, etc.) to hold atomic particles together and make the atom work. It now appears that none of these are necessary.

We have been taught that atomic shells are elliptically flat, can hold more than two electrons, and that each electron within a shell is in some way different from all others. It now appears that this level of complexity is unnecessary.

As Newton's gravitational constant (G) is based upon Quanta why shouldn't his theories also apply to atoms?

We have been advised by the world's most eminent astrophysicists that it is impossible to calculate spin in satellites and force-centres. Yet Newton's laws provide us with all the information needed to solve this problem.

Nobody appears to have grasped the fact that Newton's formula directly (with no reinterpretation) allows us to calculate the pressure inside a solid body, such as a planet or star, so why are we still guessing internal pressures?

In fact, guesswork appears to be prevalent throughout science today.

Together with the help of a number of early heroes (refer to Appendix A8), Newton provided everything we need to understand our universe ...
... how it was created,
... the age of everything in it,
... how it works,
... what everything in it is made of,
... how it generates its energy and
... where it stores this energy;

The end of the 18th century saw the start of the industrial revolution, which continues today, only now; it is called a technological revolution. The start of the 20th century should have kicked off a scientific revolution. It never happened. Why?

1.1.1 The Photon

The problem was the photon.

I need to deal with this issue now in order that it doesn't interfere with your understanding of the universal model discussed in this book.

It is about time we all dropped the concept of photons, i.e. the belief that electrons travelling at the speed of light emit light; *they don't*.
We have been taught this for a hundred years, forcing us to create weird and wonderful theories to explain how *mass* moves in waves; *it doesn't*.
The photon exists in our minds because of a very simple mistake made a long time ago related to Crooke's tube (refer to Chapter 6.4).

Once this is understood, the whole problem of energy, magnetism, gravity, electricity, etc. vanishes. You can ignore quantum theory and the theory of relativity, both of which were invented to explain this misunderstood behaviour of electrons.

The deflection of light can *only* be explained using Newton's gravitational constant (G), and the behaviour of electrons within atoms can *only* be resolved using Newton's laws of orbital motion and Coulomb's laws of electrical force (refer to Chapters 6.2.1 & 6.11.2). We should not, however, forget William Gilbert's contribution, which predates and forms the basis of all the theories related to force and energy fields (both atomic and astronomic).

It appears to me that if everybody had realised that Crooke could not possibly have created a perfect vacuum in his tube, we would not have been confused by quantum theory and the theories of relativity, and we would now be 100 years into a '*scientific revolution*'.

1.2　And Now?

Whilst the theories proposed in this book concerning Newton's Laws of Orbital Motion, Orbital Systems, Planetary Spin, Core Pressure, the Atom and Earth's Magnetic Field are a matter of scientific fact, those on Energy and the universe are hypotheses.

However, they ...
... are based on and obey well-known universal laws of nature that work
... have no need for statistics, unification theories or obscure concepts
... reflect what we sense in the universe
... have no need for intimidation

It cannot have escaped everyone's notice that Newton's, Coulomb's, Gilbert's, Maxwell's and others' force formulas all have the same configuration:
$F = K.v_1.v_2 / R^2$ (which is actually: $F = K.v_1.v_2 / A$)
where: 'K' is a constant, 'v' a variable and 'A' the spherical surface area at radius (R).

My own calculations have revealed a similar relationship for the conversion of electro-magnetic energy to velocity in electrons:
$T = X.v^2 / e^2$
where 'X' is a constant, 'v' the velocity of an electron and 'e' its electrical charge.

If all these formulas *look* the same, they probably *are* the same, i.e. they are simply variations based upon our current misunderstanding of gravity, mass, heat, etc. which are actually the same thing; *energy*. Thus, there are really only two formulas, one of which is for electrical force (Coulomb) and the other for magnetic force (Gilbert/Newton) that differ by a coupling ratio (φ = 4.407E-40). Given that gravity is magnetism, we need be in no doubt that Newton's formula represents magnetic force and can be explained as such (refer to Chapter 6.8).

The atomic model proposed here is elegant, eternal, predictable and brilliantly simple; anyone can understand it without the need for shock-tactics. It also complies with all of Newton's, Gilbert's, Coulomb's, Faraday's and Maxwell's laws, so there is no need for unification theories or statistics. In fact, it now looks highly likely that contrary to popular scientific opinion, these laws are sufficient to explain everything in our universe. Newton's laws are indeed universal, and via them, we can create realistic solutions for virtually everything in our universe from atomic to astronomic physics, including: neutronic energy, '*Big-Bang*', Earth's magnetic field, 'G', 'E=mc²', ultimate density and a great deal more.

Everything is energy: our universe is very much simpler than the one we have been taught, and exploited properly it can provide us with all the clean, free energy we need, simply from Newton's orbits.

If my model is correct (or even close), it then becomes a simple, albeit time-consuming enterprise to determine everything there is to know about our universe, from the very smallest (Quanta) to the very largest (*Big-Bang*) using theories that have been known since Poincaré first revealed his formula and Crooke discovered electro-magnetic energy in the 19th century.

1.3   Where Do We Go From Here?

Given what we now know about universal energy;
1) How it is created (orbits and spin-friction)
2) Where it is created (stars and planets)
3) How it is transmitted (electro-magnetic energy)
4) Where it is stored (neutrons)

We now have access to unlimited, clean, free energy sources;
1) Elliptical Orbits
2) Mantle heat
3) Neutrons

Moreover, these theories can give us the ability to *mathematically* predict chemical reactions in *all* matter irrespective of complexity; the *ultimate calculator*.

Such a calculator would preclude the need for material, chemical or pharmaceutical testing and experimentation. No more risk, material, time or money need be wasted on such activities and every country in the world would be able to design [100% accurate] new materials, chemicals and medicines in safety, from a computer terminal with trained but semi-skilled personnel. Furthermore, the creation of comprehensive organic and inorganic chemical databases will remove the need for duplicate effort together with the horrendous qualification periods for new medicines imposed by various national and international health authorities.

Because we now know where the universe stores its energy, we have access to an unlimited supply free from waste and pollution. We could do something useful with the world's nuclear waste; as the fuel for clean, controllable, efficient energy generators of any size. Much less mining!

Moreover, due to the discovery of the true meaning of $E=mc^2$ (refer to Chapter 6.2.5), there is no longer any reason to assume that light-speed is a limiting condition for matter. And if matter has no mass, imposing a limiting velocity owing to the conversion of mass to energy becomes unnecessary. The speed of light is simply a speed for electro-magnetic radiation, such as that for sound: there's no reason it cannot be exceeded.

Anti-gravity also becomes *theoretically* possible. All you need to do is repel the earth's *magnetism*, which is easier than opposing *gravity* with mass.

A few of the possibilities from the discoveries explained in this book are listed below?

1) Molecular calculator (and database) giving new (perfect) materials, medicines and chemicals in minutes
2) Clean, free efficient energy (by-product = hydrogen)
3) Propulsion-free satellites
4) The ability to safely recycle nuclear waste
5) Energy cells that can be fuelled with any matter (e.g. rocks!)
6) Alter elements into something else
7) Change the colour of matter electrically
8) Together with PERS#, the elimination of skin-friction offers virtually free travel
9) Perfect lubricants (machines with almost eternal life)
10) Free energy from the earth's mantle
11) Massive reductions in: pollution, material waste, energy, etc.
PERS = potential energy recovery system

In other words, we now have the ability to …
… massively reduce energy and battery production;
… massively reduce mining requirements;
… massively reduce transport costs;
… massively reduce the number of chemical laboratories;
… eliminate; national power stations & transmission lines, wind-turbines & solar panels;
… eliminate pollution from energy generation;
… create vehicles with no engine or drivetrain that need no refuelling;
… create 100% recyclable packaging

All the energy we use today requires the generation of much more to harness and recycle it. Instead of generating energy at an efficiency of less than 10%, we now have access to energy generation that is 231,000,000% efficient.

Instead of swapping one pollutant for another and/or simply moving it around as we do today, we could now create a genuinely clean place for everyone in which to live; together with limitless cheap energy for all.

1.4 How This Book Is Organised

This volume provides a non-technical description of this universal model:

2 Narrative

A written description that gives a general overview of the various discoveries made in this book. It is devoid of formulas and mathematical complexity with a view to providing a *'light-read'*!

3 Calculation Procedures (Vol II)

A compilation of the mathematical formulas supporting the narrative, including how to use them. This section has been written to simplify their use.

4 Calculation Results (Vol II)

A collection of [mostly] tabulated calculation results for selected examples using the formulas provided in section 3 (above).

5 Physical Constants (Vol III)

All the physical constants (including electrical properties such as Volts, Amps, Henries, Farads, Ohms, etc.) are provided (to ≤15 decimal places) in terms of the same four basic units; length, time, mass and charge and two ratios: m_e, e, R_n, t_n & ξ_v, ξ_m

6 Support (Vol III)

A mathematical and descriptive explanation for all the physical constants and scientific discoveries along with the reasons why Relativity and Quantum Theory must now be discarded.

7 Things You Can Do! (Vol III)

A list of unresolved issues.

8 Appendices

References, symbols, glossary, etc. used throughout this book along with a summary list of corollaries and hypotheses.

For reference purposes, the contents List and page numbering in volumes I, II & III match the page numbering in the complete book.

5 The Physical Constants

All the physical constants (including electrical properties such as Volts, Amps, Henries, Farads, Ohms, etc.) are provided (to ≤15 decimal places) in terms of the same four basic units; length, time, mass and charge and two ratios: m_e, e, R_n, t_n & ξ_v, ξ_m

5.1   Introduction

Since venturing into this mathematical field almost three years ago, what struck me most, was the inability to define mechanical, electrical and magnetic properties in terms of the same units. It is inconvenient to define mechanical properties in terms of electrical and/or magnetic energy.

Formulas for permittivity, magnetic constant, etc. are far too obscure and Planck's universal energy constant (h) actually has incorrect units. The magnetic field constant (B) is simply the reciprocal of the relative charge capacity (RC) and therefore becomes redundant. Whilst heat capacity coefficients exist, there are no such coefficients for charge. Etc.

Every document I have used in the past offers approximate values for all constants. Some of the more overt publications actually add ±Tolerance values in brackets to show how clever they are! In fact, there are only a few (7) primary constants, the values for which are known accurately. As all the others can be calculated, it is a simple matter to establish accurate values for *all* constants. We shouldn't need to qualify them with '(±??)'. For this reason, I have provided values for all constants accurate to 15 significant figures, except where absolute accuracy requires less. I leave it up to you to round them off if preferred. Approximations are unnecessary.

In the light of my discoveries, not least the neutronic radius (R_n), I have defined the properties of *all* astronomic and atomic forces and energies in these same universal units, making it much easier to understand the connection between the orbital systems.

I do not claim that everything in this publication is true and exact, merely that it represents the best and most useable collection of natural constants compiled to date. Please let me know of any errors or omission, and I shall update this publication accordingly.

The physical constants are the most important part of evaluating and defining natural laws, all of which are calculated here from just five fundamental constants (mass, charge, length, time & temperature) and two ratios (mass and velocity)

Because everything in the universe is energy, and we have (to date) concentrated on defining nature in terms of mechanics, the true nature of our universe has become very difficult to reconcile.

For example; we cannot readily explain Volts, Amps, Ohms, etc. in terms of energy. Therefore, I have reduced everything to the same basic [metric] units (Imperial conversions are in parenthesis):

Mass: kilogram (kg)
(1kg = 2.204622621lb)

Length: metre (m)
(1m = 39.3700787401575in)

Time: second (s)
(1s = 1s)

Electricity: Coulomb (C)
(1C = 1C)

Temperature: Kelvin (K)
(1K = 1.8 R)

These are the fundamental units that define all others. *Everything* can be explained and described (mathematically) using them.
Temperature is only a form of measurement for electron kinetic energy; it is not required to explain any natural property.
The only constant not fully resolved is the unit of magnetic charge, which is currently described as mass (kg).

Notes:
1) Joules and Newtons remain useful, but they are merely compilations of the above.
2) Converting to imperial units ...
... between numerators or denominators: multiply by the conversion factor above
... across numerators and denominators: divide by the conversion factor above

5.2   Symbols

The following is an alphabetical list of the symbols explained in the Tables (in this Chapter 5) indicated; **Table 5.?**.

Those that are new, i.e. not currently known, are highlighted in **bold text**

Symbol	Description	Table(s)
a_0	Rydberg radius (also known as Bohr Radius)	4
A	electrical current	8
c	speed of electro-magnetic radiation	4
$C_?$	specific heat capacity	6
$C_?$	heat capacity	5, 6
e	elementary charge unit	3
e	natural logarithm	4
E	energy	6.10.1 & 2, 6.10
F	Farad	8
F	Force	6.10.1 & 2, 6.10
G	Newton's gravitational constant	4
h	Planck's constant	4
ℏ	Planck's constant (Dirac version)	4
h'	modified Planck's constant	4
H	Henry	8
k	Coulomb's constant	4
k'	Coulomb's constant (modified)	4
k_B	Boltzmann's constant	5
K	Constant of proportionality	Chapter 6.11.14
m	mass	6.10.1 & 2, 6.10
m_e	mass of an electron	3
m_p	mass of a proton	4
m_n	mass of a neutron	4
$N_?$	microstate	5
N_A	Avogadro's number	6.11.12
q	specific charge capacity	7
Q	charge capacity	5, 7

Table 5.1a

Symbol	Description	Table(s)
r	particle (or body) radius	6.10.1 & 2
R	orbital radius	4.5.1 to 4
$R_?$	gas constant	5
R_a	specific gas constant	6
RAC	relative atomic charge	7
RAM	relative atomic mass	6
R_c	charge [emission] capacity	5, 7
RC	relative charge capacity	5
R_i	ideal gas constant	5
R_n	neutronic radius	4
R_p	relative charge capacity (constant pressure)	5
R_T	gas constant (temperature dependent)	5
R_∞	Rydberg's wave number	4
R_γ	Rydberg's universal constant (energy)	4
t	time	6.10.1 & 2, 6.10
t_n	neutronic period	3
$T_?$	Temperature (key)	4.5.1 to 4
T_n	neutronic temperature	3
v	velocity	4.5.1 to 4
V	electrical voltage	8
V	volume	6.10.1 & 2
X	heat coefficient (velocity)	5
X_R	heat coefficient (orbital radius)	5
Y	temperature coefficient	4
ε_o	permittivity of a vacuum	4
λ	wavelength	6.10.1 & 2, 6.10
μ', μ_o	magnetic constant	4
ρ	density	6.10.1 & 2
Σ	universal constant	3
φ	coupling ratio	4
Ω	electrical resistance	8
ξ_m	mass ratio	4
ξ_v	velocity ratio	4

Table 5.1b

Symbol	Description
Suffix:	
e	electron
n	neutronic
n	atomic shell number
p	proton
p	constant pressure (heat & charge capacity)
t	constant temperature (heat & charge capacity)
u	ultimate
v	constant volume (heat & charge capacity)
Atomic: temperature, velocity & orbital radius:	
c	cold
m	mean Planck value
n	neutronic
o	minimum Planck value
Modifier:	
N	Newton
P	Planck

Table 5.2c

5.3 Primary Constants

The Tables in this Chapter provide the meaning of the symbols previously listed in Chapter 5.2. In the following Chapters, the constants are grouped according to their properties; general, heat, charge, etc.

There are very few **primary constants**, i.e. those that we must take for granted and on which *all* others are based; these are listed below

Symbol	Value	Table(s)
m_e	**9.1093897E-31**	kg
The mass of an electron (refer to Chapter 6.7)		
R_n	**2.81793795383896E-15**	m
The neutronic radius (refer to Chapter 6.11.10)		
e	**1.60217648753E-19**	C
Elementary charge unit (refer to Chapter 6.11.6)		
t_n	**5.90596121302193E-23**	s
Neutronic period		
T_n	**623316124.717178**	K
Neutronic Temperature		
ξ_m	**1836.15115053207**	
The mass ratio $\{m_p/m_e\}$		
ξ_v	**1722.0458764934**	
The velocity ratio $\{c/v_o\}$ (refer to Chapter 6.11.5)		
Σ	**3E-91** (exact)	m^6
Universal constant (refer to Chapter 6.11.1)		
Table 5.3		

Important Note: *The magnetic field constant (B) is not included here as it is simply the reciprocal of the relative Charge Capacity (RC: Table 5). When using 'B' in a formula, such as Lorentz's magnetic force formula, remember: B = 1/RC*

New properties, i.e. those not currently known or used, are highlighted in the following Tables in **bold text**.

5.4   General Physical Constants

Symbol	Formula	Value	Units
G	$a_o.c^2 / \rho_u$	6.67359232004334E-11	m³ / s².kg
Newton's gravitational constant (per m³)			
k	$c^2.\mu'$	8.98755184732667E+09	J.m / C²
Coulomb's constant [for an electron] (refer to Chapter 6.11.4)			
k'	$k / \xi_m{}^2$	2.6657815048876E+03	J.m / C²
Coulomb's constant for a proton			
φ	$G.m_e.m_p / k.e^2$	4.40742111792334E-40	
Coupling Ratio			
μ'	$\mathbf{R_n}.m_e/e^2$	1E-07	kg.m / C²
Magnetic constant (fundamental)			
μ_o	$4\pi.\mu'$	1.25663706143592E-06	kg.m / C²
Magnetic constant			
ε_o	$1 / \mu_o.c^2$	8.85418775855161E-12	C² / J.m
Permittivity of a vacuum (e.g. within an atom)			
h	$\tfrac{1}{2}.\mathbf{R_n}.m_e.c.\xi_v$	6.62607174469163E-34	kg.m²/s
Planck's constant (resolved into its component parts)			
ħ	$h / 2\pi$	1.05457207144921E-34	kg.m²/s
Planck's constant (modified by Dirac)			
h'	$\tfrac{1}{2}.\mathbf{R_n}.m_e.c^2$	1.15353857232684E-28	J.m
Modified Planck's constant			
R_∞	$1 / a_o.\xi_v$	1.09737269561359E+07	/m
Rydberg's wave number			
R_y	$\mathbf{R_n}/a_o . \tfrac{1}{2}.m_e.c^2$	2.17987197684936E-18	J
Rydberg's universal constant for the energy of an electron			
α	$e^2 / 4\pi$	2.04272942122269E-39	C²
Fine structure constant			
X	$\mathbf{T_n}/c^2$	6.9353271647894E-09	K.s²/m²
Velocity constant			
X_R	$\mathbf{T_n}.R_n$	1.75646616508035E-06	K.m
Radial constant			
Y	$\sqrt[3]{[\tfrac{1}{2}.\xi_v]}$	9.51345439232503	
Temperature coefficient			
e'	$e.\xi_v.\sqrt{[T/T_n]}$	2.75902141376572E-16	
Proton charge			
Table 5.4a			

Atomic property constants (refer to Chapter 6.10 for particle properties):

Symbol	Formula	Value	Units
m_e			kg
Mass of an electron (refer to Chapter 5.3)			
m_p	$m_e.\xi_m$	1.672621637830E-27	kg
Mass of a proton			
m_n	m_e+m_p	1.6735325768E-27	kg
Mass of a neutron			
a_o	$R_n.(\xi_v/4\pi)^2$	5.2917721067E-11	m
Rydberg's radius			
R_c	$R_n.\xi_v^3$	1.43901585166681E-05	m
Cold orbital radius (refer to Chapter 3.3.4)			
R_o	$R_n.\xi_v^2$	8.3564315638157E-09	m
Planck minimum orbital radius (refer to Chapter 3.3.4)			
R_m	$R_n.\xi_v$	4.85261843362263E-12	m
Planck mean orbital radius (refer to Chapter 3.3.4)			
R_n			m
Neutronic radius (refer to Chapter 3.3.4)			
v_c	$v_o . \sqrt{[R_o/R_c]}$	4195.20925990715	m/s
Electron cold velocity (refer to Chapter 3.3.4)			
v_o	$c . \sqrt{[R_n/R_o]}$	174090.866621084	m/s
Electron minimum Planck orbital velocity (refer to Chapter 3.3.4)			
v_m	$\sqrt{[c.v_o]}$	7224342.80705004	m/s
Electron mean Planck orbital velocity (refer to Chapter 3.3.4)			
c	$2\pi.R_n / t_n$	299792459	m/s
Electron neutronic velocity (refer to Chapter 3.3.4)			
T_c	$X.v_c^2$	0.122060237421696	K
Cold temperature (refer to Chapter 3.3.4)			
T_o	$X.v_o^2$	210.193328535837	K
Planck minimum temperature (refer to Chapter 3.3.4)			
T_m	$X.v_m^2$	361962.554671561	K
Planck mean temperature (refer to Chapter 3.3.4)			
T_n	$X.c^2$		K
Neutronic temperature (refer to Chapter 3.3.4)			
e	exp(1)	2.71828182845905	
Natural logarithm			

Table 5.4b

5.5 Universal Heat & Charge Capacities

Symbol	Formula	Value	Units
k_B	$m_e.c^2 / Y.T_n$	1.38065156E-23	J/K
Boltzmann's constant (refer to Chapter 6.11.13)			
R_i	$k_B.N_A$	8.31447876657891	J / K.mol
Ideal gas constant			
RC	e/m_e	1.75881869180545E+11	C/kg
Relative charge capacity			
R$_c$	$\sqrt{[\,G/k\,]}$	8.61706029887134E-11	C/kg
Charge [emission] capacity			
R_a	R_i / RAM		J / kg.K
Specific gas constant			
R	$m.R_a$	1.38065156E-23	J/K
Gas constant			
R$_p$	c_p.RAM	20.7861969164473	J / K.mol
Gas constant; R_i multiplied by 2.5			
R$_T$	$RAC.q_p.Ln(T)$ $R_i.Ln(N_t)$		J / K.mol
Gas constant ($R_T = R_i$ when $N_t = e$ & $T = 1.49182469764127$ K)			
C_t	$m.c_t$		J/K
Heat capacity (constant *temperature*)			
C_v	$m.c_v$		J/K
Heat capacity (constant *volume*); C_t multiplied by 1.5			
C_p	$m.c_p$		J/K
Heat capacity (constant *pressure*); C_t multiplied by 2.5			
Q$_t$	$e.\mathbf{q_t}$		J/K
Charge capacity (constant *temperature*); also equal to **R** & C_t			
Q$_v$	$e.\mathbf{q_v}$		J/K
Charge capacity (constant *volume*); **Q$_t$** multiplied by 1.5; also equal to C_v			
Q$_p$	$e.\mathbf{q_p}$		J/K
Charge capacity (constant *pressure*); **Q$_t$** multiplied by 2.5; also equal to C_p			
Table 5.5a			

5.5.1   Microstates

Symbol	Formula	Value	Units
N_t	$\exp(c_p.L_n(\underline{T}) / R_a)$ $\exp(\mathbf{q_p}.L_n(\underline{T}) / R_a)$ $\exp(2.5 . Ln(\underline{T}))$		
Microstate (constant *temperature*)			
N_V	c_v / R_a $\mathbf{q_v} / R_a$		
Microstate (constant *volume*); N_t multiplied by 1.5			
N_p	c_p / R_a $\mathbf{q_p} / R_a$		
Microstate (constant *pressure*); N_t multiplied by 2.5			

Table 5.5b

5.6 Specific Heat Capacities (particles)

Symbol	Formula	Value	Units
RAM$_e$	$m_e.N_A$	5.4858031839070700E-07	kg/mol
Relative atomic mass of an electron			
RAM$_p$	$m_p.N_A$	1.00727638277235E-03	kg/mol
Relative atomic mass of a proton (also the RAM of an hydrogen atom)			
R_{ae}	R_i / RAM$_e$ = k_B / m_e	1.51563563034308E+07	J / kg.K
Specific gas constant for an electron			
R_{ap}	R_i / RAM$_p$ = k_B / m_p	8.25441647276088E+03	J / kg.K
Specific gas constant for a proton			
c_{et}	k_B / m_e	1.51563563034305E+07	J / kg.K
Specific heat capacity for the electron (constant *temperature*)			
c_{eV}	1.5 . c_{et}	2.27345344551458E+07	J / kg.K
Specific heat capacity for the electron (constant *volume*)			
c_{ep}	c_{et} + c_{eV}	3.78908907585763E+07	J / kg.K
Specific heat capacity for the electron (constant *pressure*)			
c_{pt}	k_B / m_p	8.25441647276074E+03	J / kg.K
Specific heat capacity for the proton (constant *temperature*)			
c_{pV}	1.5 . c_{pt}	1.23816247091411E+04	J / kg.K
Specific heat capacity for the proton (constant *volume*)			
c_{pp}	c_{pt} + c_{pV}	2.06360411819018E+04	J / kg.K
Specific heat capacity for the proton (constant *pressure*)			
C_t	$m_e.c_{et}$ $m_p.c_{pt}$	1.38065156E-23	J/K
Heat capacity (constant *temperature*); equal to **R** & **Q$_t$**			
C_V	$m_e.c_{eV}$ $m_p.c_{pV}$	2.07097734E-23	J/K
Heat capacity (constant *volume*); equal to **Q$_V$**			
C_p	$m_e.c_{ep}$ $m_p.c_{pp}$	3.4516289E-23	J/K
Heat capacity (constant *pressure*); equal to **Q$_p$**			

Table 5.6

5.7   Specific Charge Capacities (particles)

Symbol	Formula	Value	Units
RAC_e	$e.N_A$	96485.3317942158	C/mol
Relative atomic charge of an electron (also equal to the Farad)			
RAC_p	$e'.N_A$	1.77161652983418E+08	C/mol
Relative atomic charge of a proton (also the RAC of an hydrogen atom)			
R_{ce}	R_i / RAC_e	8.61735002820125E-05	J / C.K
Specific gas constant for an electron			
R_{cp}	R_i / RAC_p	4.69315939796359E-08	J / C.K
Specific gas constant for a proton			
q_{et}	k_B / m_e	8.61735002820123E-05	J / C.K
Specific charge capacity for the electron (constant *temperature*)			
q_{ev}	$1.5 . q_{et}$	1.29260250423019E-04	J / C.K
Specific charge capacity for the electron (constant *volume*)			
q_{ep}	$q_{et} + q_{ev}$	2.1543375070503E-04	J / C.K
Specific charge capacity for the electron (constant *pressure*)			
q_{pt}	k_B / e'	4.69315939796358E-08	J / C.K
Specific charge capacity for the proton (constant *temperature*)			
q_{pv}	$1.5 . q_{pt}$	7.0397390969454E-08	J / C.K
Specific charge capacity for the proton (constant *volume*)			
q_{pp}	$q_{pt} + q_{pv}$	1.1732898494909E-07	J / C.K
Specific charge capacity for the proton (constant *pressure*)			
Q_t	$e.q_{et}$ $e'.q_{pt}$	1.38065156E-23	J/K
Charge capacity (constant *temperature*); equal to **R** & C_t			
Q_v	$e.q_{ev}$ $e'.q_{pv}$	2.07097734E-23	J/K
Charge capacity (constant *volume*); equal to C_v			
Q_p	$e.q_{ep}$ $e'.q_{pp}$	3.4516289E-23	J/K
Charge capacity (constant *pressure*); equal to C_p			
Table 5.7			

5.8   Electricity

Apart from the Farad, no values are provided for the following electrical properties because the Amp, Volt, Ohm and Henry are now redundant.

Symbol	Formula	Value	Units
A	$e.f$		C/s
Electrical current (Coulomb flow-rate)			
V	PE/e		J/C
Electrical voltage (potential energy per coulomb)			
Ω	$V/A = PE / f.e^2$		$J.s/C^2$
Electrical resistance (momentum over distance per Coulomb squared)			
H	$\mu_o.R$		$kg.m^2/C^2$
Henry; unit of mutual inductance			
F	$e.N_A$	96485.3317942156	C/mol
Farad; unit of electrostatic capacitance (equal to **RAC$_e$**)			
P	$V.A = PE.f$		J/s
Power (Watt)			

Table 5.8

PE is the potential energy between a proton and its orbiting electron

6 Support

A mathematical and descriptive explanation for all the physical constants and scientific discoveries along with the reasons why Relativity and Quantum Theory must now be discarded.

6.1   Proof of the Orbital Model

The four principal agents for the theories of planetary motion were Copernicus, Kepler, Galileo and Newton. Between them, they defined the behaviour of orbiting satellites, moons and planets that remain valid even today.

6.1.1   Nicolaus Copernicus (1473 to 1543)

Copernicus stated that; contrary to religious doctrine, the sun does not orbit the earth, but all the planets in the solar system orbit the sun. He was so concerned for his safety regarding this claim, however, that he arranged for the publication of his findings to be deferred until after his death (1543).

6.1.2   Johannes Kepler (1571 to 1630)

Kepler used Tycho Brahe's (1546 to 1601) observational data to show that the planets not only orbited the sun, just as Copernicus had previously claimed, but that their orbital paths were ellipses. Kepler also stated that the time taken to traverse between any two points (refer to Chapter 2.2.2; Fig 5) on this elliptical curve is proportional to the swept area:

i.e; $t_1/A_1 = t_2/A_2$

Whilst he did not provide a mathematical proof for his swept area theory, he understood it. It was later confirmed by Isaac Newton (below).

6.1.3   Galilei Galileo (1564 to 1642)

Galileo is best known for his physical evidence of celestial bodies (moons) orbiting other planets, revealed in his book; Dialogue Concerning the Two Chief World Systems (frequently referred to as the 'Dialogue'), therein declaring Copernicus correct and finally quashing over a thousand years of religious dogma that stated all celestial bodies orbit the earth. In return for his findings, he was put under permanent house arrest, but only after being threatened with death if he didn't recant this claim.

However, it was during his confinement that Galileo completed his most important work, his laws of motion, one of which states that a body fired from the surface of the earth would follow a parabolic curve back to its surface.

This claim may be demonstrated by comparing Galileo's mathematically correct parabola with a projectile trajectory calculation:

$x(t) = A.t + B$

$y(t) = C.t + D - \frac{1}{2}.g.t^2$

If B and D are zero {i.e. v occurs at t = 0}:

$x(t) = v.Cos(\alpha).t$

$y(t) = v.Sin(\alpha).t - \frac{1}{2}.g.t^2$

Where:
v = initial velocity
A = initial horizontal velocity {i.e.; A = v.Cos(α)}
B = offset horizontal distance from t = 0
C = initial vertical velocity {i.e.; C = v.Sin(α)}
D = offset vertical distance from t = 0

Fig 34 shows the projectile trajectory (curve) superimposed on two alternative parabolic curves, one of which passes through the same latus rectum and the other being the best parallel match.

Whilst the parabolic path is not strictly correct, it is stunningly close, demonstrating that given the limited information and facilities available to Galileo at his time, he was a very capable mathematician.

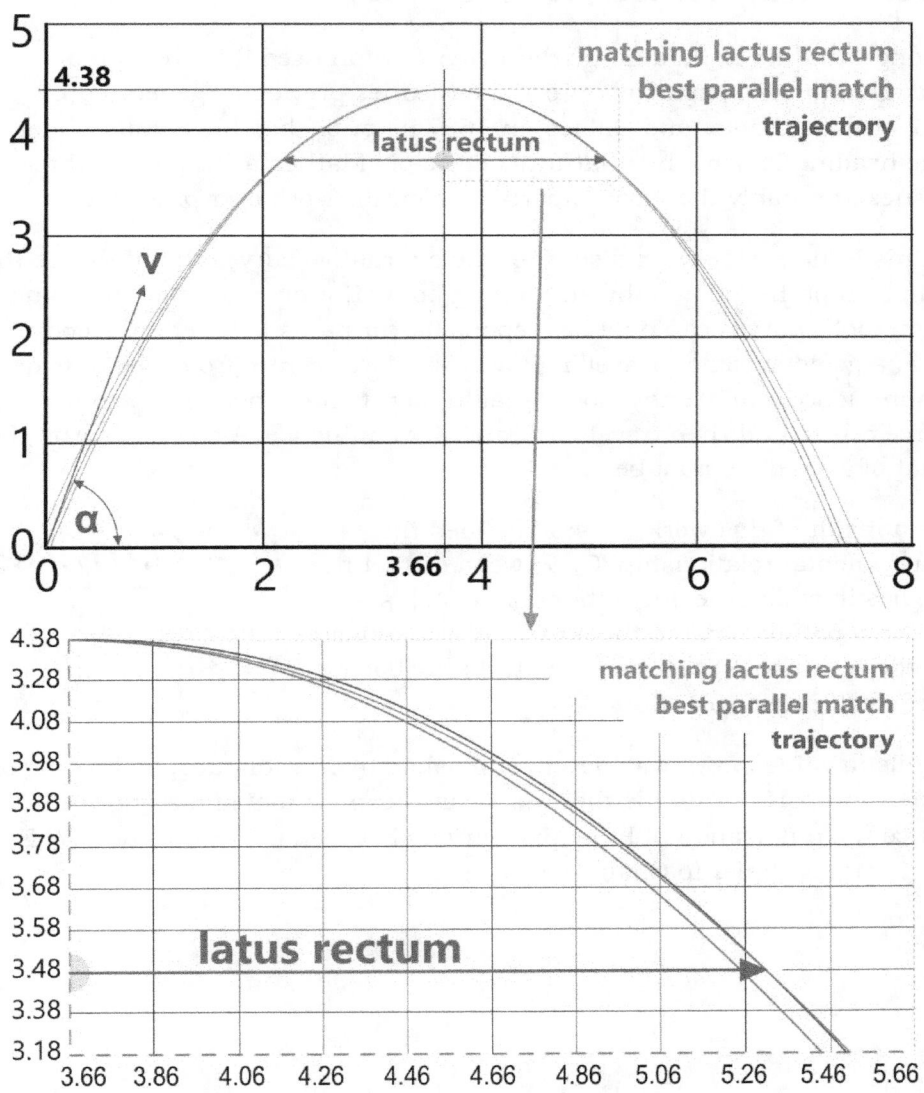

Fig 34

6.1.4   Isaac Newton (1642 to 1727)

Along with [his explanation for] gravity, Newton used [his creation of] calculus to mathematically prove the theories previously generated by Copernicus, Kepler and Galileo. In 1687 he published his results under the heading Philosophiæ Naturalis Principia Mathematica (first of three issues), probably the most important scientific work ever produced.

In his Principia, Newton discusses the alternative curves that describe the elliptical paths followed by an orbiting body. However, the parabolic and hyperbolic curves can only be responsible for paths followed by a body (e.g. a galactic comet) travelling towards a force centre from well outside its influence, sufficiently close to fall under its influence, pass around the force centre and then travel back out of its influence. A complete orbit, i.e. that of a satellite must be an ellipse.

As a result of this work, Newton defined the fundamental relationship (G) between attracting bodies in which the gravitational force (F) is directly proportional to the inverse of the square of the distance (R) between the attracting bodies (refer to Chapter 6.1.4.6).

$$F = \frac{G.m_1.m_2}{r^2}$$

Fig 35

Whilst a value for 'G' was never established by Newton, despite it being of special importance to his theories, it has been estimated many times since the publication of Principia, varying between 6.67E-11 and 6.76E-11 N.m²/kg² (refer to Chapter 6.11.2)

The minimum and maximum radial distances between the earth and sun (Fig 36: @ A & B respectively) are assumed to be as defined in the Earth-Sky fact sheet (https://earthsky.org/). Therefore, using Newton's theories and true value for 'G' (refer to Chapter 6.11.2), the principal properties of the earth's orbit are as follows:

a = 1.495945981E+11m (R + R)/2)
b = 1.495737135E+11 {√[a².(1-e²)]}
e = 0.01670914665 {a.e² + R.e + R - a = 0}
p = 1.495528319E+11m {a.(1-e²)}
f = 1.47095E+11m {a.(1-e)}
x' = 2.499598078E+09m {a-f}
R = 1.47095E+11m to 1.520941962E+11m
F = 3.658178805E+22 to 3.421649078E+22N (refer to Chapter 6.1.4.6)
v = 30286.008788376 to 29290.53557m/s (refer to Chapter 6.1.4.9)

Newton's creation of Calculus allowed him to generate formulas for non-linear versions of Galileo's relationships for distance (s), time (t), velocity (v) and acceleration (a) as follows:

s = ut + ½at²
δs/δt = v = u + at
δ²s/δt² = δv/δt = a

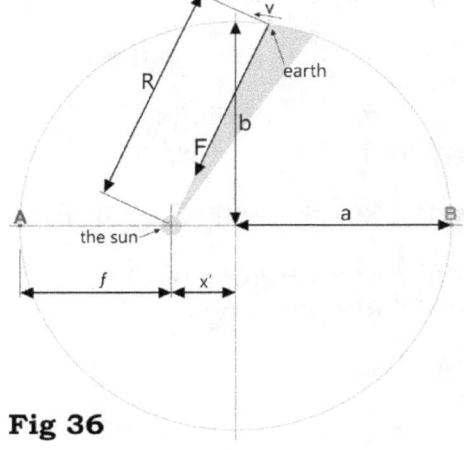

Fig 36

6.1.4.1 Proof (elliptical orbits)

By applying calculus, Newton was able to generate the non-linear formulas necessary to complete his theories concerning the elliptical (conic) path of orbiting bodies, which was proven as follows:

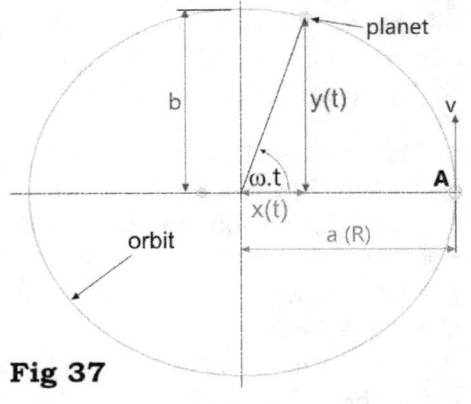

Assume an ellipse and the planet is passing the x-axis @ 'A' (y = 0)
(Fig 37)
x component = R {a}
y component = v/ω {b}

Fig 37

$x(t) = R.Sin(\omega.t)$
$y(t) = (v/\omega).Cos(\omega.t)$

From: $Sin^2(\omega.t) + Cos^2(\omega.t) = 1$

$y(t)^2 / (v/\omega)^2 = 1 - Sin^2(\omega.t)$
$Sin^2(\omega.t) = 1 - y(t)^2 / (v/\omega)^2$

$x(t)^2 / R^2 = Sin^2(\omega.t) = 1 - y(t)^2 / (v/\omega)^2$

$x(t)^2 / R^2 = 1 - y(t)^2 / (v/\omega)^2$
$x(t)^2 / R^2 + y(t)^2 / (v/\omega)^2 = 1$

An ellipse!

6.1.4.2 Euclidean Geometry (equal areas)

Whilst Kepler had already predicted the equal-swept-area-with equal-orbital-time theory, it had still not been mathematically proven by the time Newton was writing his Principia. Newton did this using Euclidian geometry.

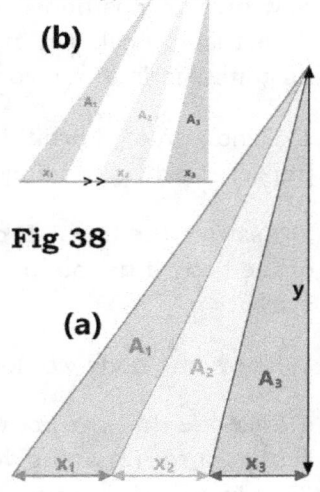

Fig 38

The areas of each triangle in Fig 38; A_1, A_2 & A_3 are all equal if the base widths; x_1, x_2 & x_3 are equal, which can be proven as follows:

Let $y=6$ and x_1, x_2 & x_3 all equal 3 (a)
The area of a triangle: $A = x.y/2$
$A_1 = 6 \times 3 \div 2 = 9$
$A_2 = 6 \times (3+3) \div 2 - A_1 = 9$
$A_3 = 6 \times (3+3+3) \div 2 - A_1 - A_2 = 9$
Therefore, all the areas are equal (i.e. 9)
The same applies to triangles with equal bases between parallel lines (b)

He then applied this to the conservation of energy (Fig 39)

6.1.4.3 Proof (conservation of energy & equal time-swept area)

Newton's proposition diagram for his proof of Kepler's 'equal-areas-equal-time' theory is shown in Fig 39, where the following instructions describe its construction (my words):

1) Divide time [of orbit] into equal parts [represented by equal swept areas {triangles}]

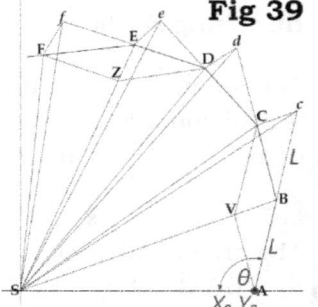

Fig 39

2) Assume the line A-B describes the linear path of the body if unconstrained by gravitational attraction

3) The same body would then continue to B-c

4) Assume that the body is attracted by a central-force (S) and diverted from its right line (B-c) in a direction parallel to V-B as far a C

5) Continue to generate similar triangles (S-A-B) following the points D, E, F, etc.

Note: The dimensions L, θ, X_o & Y_o in Fig 39 were not part of Newton's original drawing. They have been added by me in order to assist with the correlation between Figs 39 to 41

Newton was therefore stating that all swept areas (triangles SAB, SBC, SCD, SDE, SEF, etc.) must be equal.

The difficulty in generating the above diagram is knowing how far along the line C-c that C occurs in order to ensure that each subsequent area remains equal.

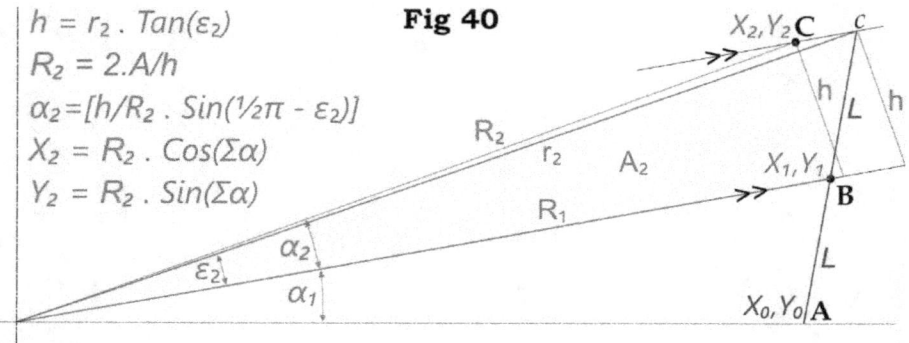

$$h = r_2 . Tan(\varepsilon_2)$$
$$R_2 = 2.A/h$$
$$\alpha_2 = [h/R_2 . Sin(\tfrac{1}{2}\pi - \varepsilon_2)]$$
$$X_2 = R_2 . Cos(\Sigma\alpha)$$
$$Y_2 = R_2 . Sin(\Sigma\alpha)$$

Fig 40

This can be achieved using the process described in Fig 40, where the blue variables are entered (X_0, Y_0, L, θ), all the green variables (X_1, Y_1, R_1, r_2, A_2, $ε_2$, X_0, a_1) can be easily calculated using the blue variables and the red variables (h, R_2, a_2, X_2, Y_2) may be determined using the formulas provided.

Newton claimed that if you reduce length L 'in infinitum' and join up the dots (X,Y co-ordinates) you produce a curved line thereby demonstrating that a centripetal force is continually acting on the body in the direction of the force centre and the triangular areas will always be proportional to the time passed by the body traversing each triangle; QED

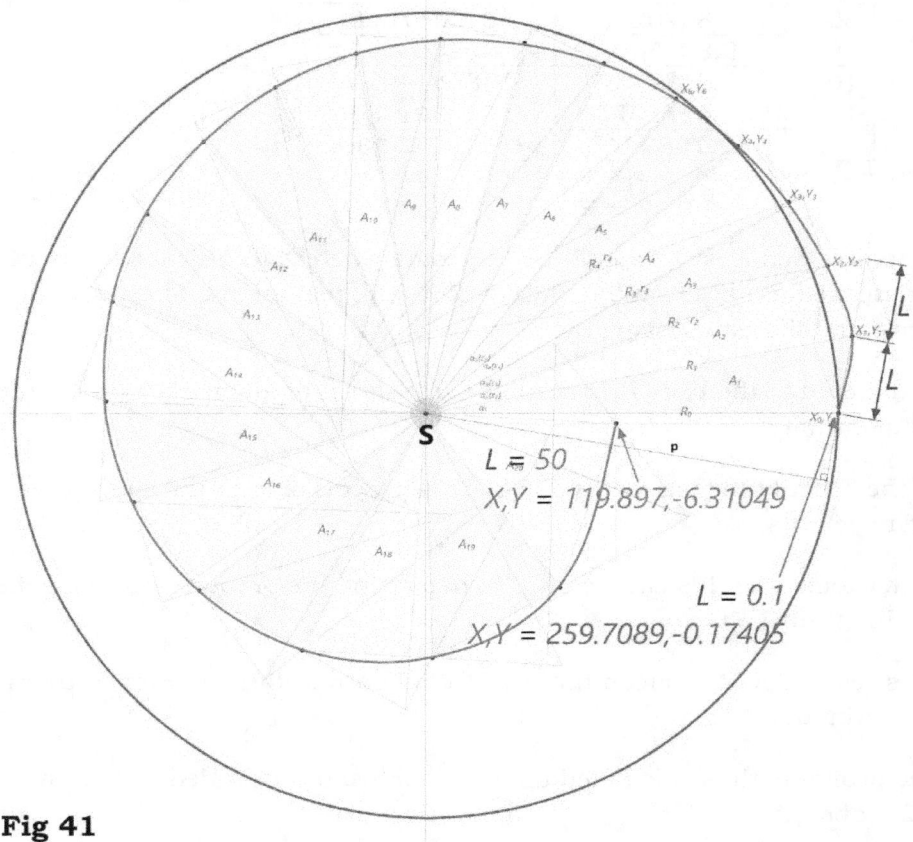

Fig 41

This argument is less easily seen from his words and his fairly simple diagram (Fig 39) than if you actually complete his diagram and repeat it for ever smaller values of L through to 360° (see below)

A calculation was carried out using the following input data:
X_0, Y_0 = **260,0**
L = 50
θ = 100°

As can be seen in Fig 41, the diagram does indeed produce a curve, exactly as Newton claimed

... and following this through a sequence of diminishing values for L from 50 to 0.1, the following X,(Y) co-ordinates are achieved immediately prior to reaching 360°:

L	X	(Y)
50	119.8970215	(-6.310487628)
25	179.800507	(-11.66860541)
10	229.5181732	(-9.486414926)
1	257.0676077	(-0.349331185)
0.1	**259.7089445**	**(-0.174048539)**

Note: the 'Y' co-ordinate is in parenthesis because it is simply a resultant. The trend is demonstrated by the 'X' co-ordinate.

... from which it isn't difficult to anticipate where X (& Y) will end up if L is diminished in infinitum [i.e.: X(,Y) = **260(,0)**], making the final shape a circle and thereby proving that:

a) the path of the body is continuous (conservation of energy and angular momentum)

b) the orbital time passed by the body is proportional to the swept area (triangle)

c) Newton's calculus can be used to determine the properties of the path {'in infinitum'}

This result does not mean that the orbital path is circular, simply that it is continuous.

The orbital path is calculated using the procedure provided in Chapter 3.2.2 above.

Corollary 1

Newton's first Corollary (to the above proof) states that the velocity of the body (v), represented by L, at positions A, B, C, D, E, F, etc. (Fig 39) is inversely proportional to the perpendicular distance of its tangent from the force centre (Fig 41; p {just to the right of "$L = 50$"})

Newton also stated that; v multiplied by p is a constant, i.e. his constant of motion (h), which is the angular momentum without the mass component.

Using the above 'in infinitum' argument it can be seen in the following table where these calculations have been carried out for successively reduced values of L between the start and end of the orbit (h_n @ 0°, h < 360°), 'h' does indeed become a constant:

L	h_n	h
50	2201.214	7485.049
25	580.0837	1196.551
10	95.12179	123.8112
1	0.963516	0.992077
0.1	**0.0970**	**0.0974**

Newton's constant of motion 'h' is not to be confused with the perpendicular distance 'h' shown in Fig 40; they are neither the same nor in any way connected.

6.1.4.4 Centripetal Force

Centrifugal acceleration (according to Christiaan Huygens {1629 to 1695}):
$a = R.\omega^2$
where $\omega = 2.\pi/t$
$a = \sqrt{[(R.\omega^2)^2 + (R.a)^2]}$
with constant angular momentum; $a = 0$
$a = R.\omega^2$
$a = R.(2.\pi/t)^2$

Centrifugal force:
$F = m.a$
$F = m.R.(2.\pi/t)^2 = 4.\pi^2.m\,(R/t^2)$

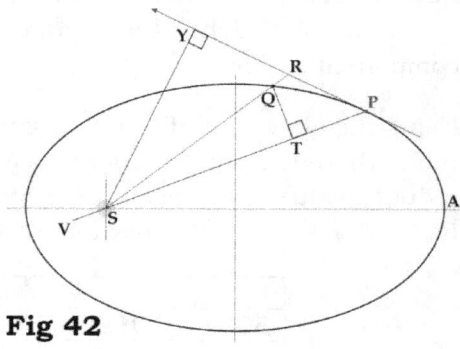

Through his inverse rules, Newton shows that the centripetal force (F) between the orbiting body and the force-centre (Fig 42);
$F = SP^2 . QT^2 / QR$

Fig 42

$PR = v^P . \delta t$
where; v^P is the velocity of the body at P and δt is the time taken for the body to travel from P to Q

$QR = (F^P / 2m).\delta t^2$
where; F^P is the centripetal force on the body at P
$F = QR . (2m / \delta t^2)$
where; F^P is the centripetal force on the body at P

$\delta t = PR/v^P = PR / (h/SY) = PR . SY / h$
where h is Newton's constant of motion (see Corollary 1 above)

Therefore, the centripetal force (F) can be calculated as follows:
$F = QR . (2.m / (PR . SY / h)^2)$
$\quad = QR . (2.m / (PR^2 . SY^2 / h^2))$
$\quad = QR . (2.m.h^2 / (PR^2 . SY^2))$
$\quad = QR.2.m.h^2 / PR^2.SY^2$

Newton preferred the calculation in geometric form by setting $2.m.h^2$ as a constant (k):
$F = k.QR / (PR.SY)^2$

6.1.4.5 Distance Between A Satellite & Its Force-Centre (R)

The separation (distance) between an orbiting body and its force centre, can be found by using general elliptical equation:

$R = a.(1-e^2) / (1-e.Cos(\theta))$

where 'R' & 'θ' are as shown in (Fig 44) and 'e' is eccentricity

The force centre is not at the centre of an ellipse but at its focus (Fig 42; S)

6.1.4.6 The Inverse Square Law

Proposition XI: *"If a body revolves in an ellipse; it is required to find the law of the centripetal force tending to the focus of the ellipse"*

Using similar geometric arguments as above (Figs 38 to 42) Newton worked out that the force between an orbiting body and its force centre is proportional to the inverse of their separation (the distance between them): $F \propto 1/R^2$ i.e. $F = K / R^2$ (refer to Chapter 2.2; Page 45)
where:
the constant of proportionality: $K = G.m_1.m_2$ i.e. $F = G.m_1.m_2 / R^2$
where G is a constant and m_1 and m_2 are the masses of the force centre and the orbiting body

This same relationship ($F \propto 1/R^2$) also applies to parabolas and hyperbolas as well as the ellipse

The above constant of proportionality (K) can also been written as;
$K = m.h^2/p$
Where 'h' and 'p' are defined in Corollary 1 above and m is the mass of the orbiting body
i.e. $F = (m.h^2/p).(1/R^2)$
In the first formula, you can resolve the problem knowing the mass of the bodies
In the second formula, you can resolve it knowing the velocity and mass of the orbiting body and the parameter of its curve (p)

Both of the above F calculations produce the same result;

e.g. the following centripetal force occurs in the earth's orbit, 0.000175° from the major semi-axis:

G = 6.67359232004332E-11 (gravitational constant)
m_1 = 1.9885E+30 (sun mass)
m_2 = 5.964519768E+24 (earth mass)
R = 1.5209420E+11 (distance between mass centres)
$F = G.m_1.m_2 / R^2$ = **3.421649078E+22** (centripetal)

h = 4.454920463E+15 (constant of motion - see Corollary 1 above)
m = 5.964519768E+24 (earth mass)
p = 1.495528319E+11 (ellipse parameter)
R = 1.5209420E+11 (distance between mass centres)
$F = m.h^2 / p.R^2$ = **3.421649078E+22**

6.1.4.7 Orbital Period

Proposition XV: "*The same things being supposed, I say, that the periodic times in ellipses are as the $3/2^{th}$ power of their greater axes*"

This means that if the major semi-axis of an ellipse is 'a' (Fig 37) and the time taken for a body to orbit the elliptical path is 't' then the relationship between the two is:

$t \propto (2.a)^{1.5}$ or $t^2 \propto (2.a)^3$

Therefore; $t = K . a^{1.5}$

Where K is the constant of proportionality, which is dependent on the properties of the force-centre.

This is actually Kepler's third law

6.1.4.8 Constant of Proportionality

To determine 'K' (the constant of proportionality for $t = K \cdot a^{1.5}$
{refer to Chapter 6.1.4.7}) ...
$K = t^2 / a^3 \ \{s^2/m^3\}$
now we know ...
... that the earth travels around the sun in 31558149s
... the earth's semi-major orbital x-axis is 1.495945981E+11m

Therefore:
t^2 / a^3 = **2.974914364E-19** $\{s^2/m^3\}$

G = 6.67359232E-11 $\{N.m^2/kg^2 = kg.m.m^2 / s^2.kg^2 = m^3 / s^2.kg\}$
m_1 = 1.9885E+30 kg (the mass of our sun)
$1 / m_1.G$ = 7.535546116E-21 $\{s^2.kg / m^3/kg = s^2/m^3\}$
2.975944645E-19 ÷ 7.538155846E-21 = 39.47841760436
$\sqrt{39.47841760436}$ = 6.2831853071796 = 2.π

Therefore:
$K = (2\pi)^2 / G.m_1 = (2\pi)^2 ÷ 6.67359232E\text{-}11 ÷ 1.9885E\text{+}30$
  = **2.974914364E-19** s^2/m^3
i.e.;
$K = (2\pi)^2 / G.m^{fc}$
where m^{fc} is the mass of the force-centre

The above calculation, based upon NASA's data for the sun and the
earth's orbit, gives an error margin of 0

6.1.4.9 Alternative Velocity Calculation

A much simpler orbital velocity
calculation method is based upon
Kepler's 'swept-area = time' rule
(Fig 43)

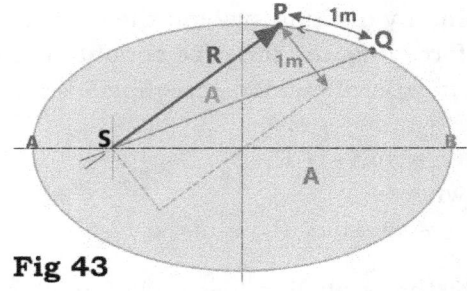

Using the earth's orbit as an
example:

Earth's total orbital area (A) is
7.029445371E+22m^2 and it takes
31558149s (t) to complete

Fig 43

The swept area (A) is equal to ½.R x 1m
The velocity of the orbiting body at any given distance between the centres
of mass (P & S) is calculated as follows:
v = 2.A / t.R {m^2 / s.m = m/s}

By way of verification:

The earth's maximum velocity occurs when R = 1.47095E+11 m (@ A)
v = (2 x 7.029445371E+22) ÷ (31558118.4 x 1.47095E+11)
 = **30286.008788376** m/s (refer to Chapter 4.2.2)
30286.008788376 m/s (calculated using; h=v.R)

The earth's minimum velocity occurs when R = 1.520941962E+11 m (@ B)
v = (2 x 7.02944537126484E+22) ÷ (31558149 x 1.520941962E+11)
 = **29290.5355716777** m/s (refer to Chapter 4.2.2)
29290.5355716777 m/s (calculated using; h=v.R)

The above confirms Kepler's 'swept-area = time' rule and shows that
v ∝ 1/R

or v = k/R
where k = 2.A / t

6.1.4.10 Centrifugal force in an orbiting body

In any orbiting system, the centripetal force, i.e. Newton's gravitational force (Fig 35), must be equal to the orbiting body's centrifugal force, which may be calculated thus (Fig 44):

$F = m_1.v_1{}^2 / R$

$Fc = m_1.v_2{}^2 / R$

where

$v_2 = \sqrt{[G.m_1 / R]}$

@ the perihelion (perigee) of an ellipse; $Fc = F . f/p = F / (1+e)$
@ the aphelion (apogee) of an ellipse; $Fc = F . p/f = F . (1+e)$

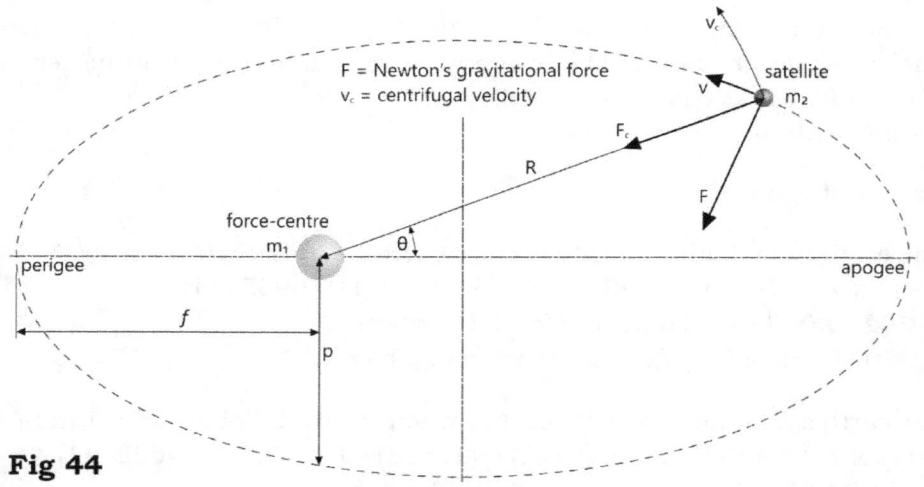

Fig 44

Orbital velocity anywhere in an orbit may be calculated thus:

$v_2 = 2\pi.a.b / R.t$

where: t is the satellite's orbital period

Centrifugal velocity anywhere in an orbit may be calculated thus:

$a = \sqrt{[{}^4/_3.\pi]}$

$\zeta = \sqrt{[(f.Sin(\theta/2)^a + p.Cos(\theta/2)^a) / (f.cos(\theta/2)^a + p.Sin(\theta/2)^a)]}$

$v_c = \zeta.v_2$

6.4.1.11 Fundamental Laws of Orbital Motion

1) Every orbital system must have a force-centre and at least one satellite

2) A force-centre's mass defines its satellite's orbital shapes and periods

3) Satellite orbits define a force-centre's spin

4) Sub-satellite orbits and force-centre spin define a satellite's spin

5) Sub-satellites have no effect on the force-centre

6) Satellites may be swapped between orbits without altering orbital shapes and periods; e.g. Jupiter may replace Earth and Jupiter would follow the same orbital path that Earth previously followed and would orbit in 365¼ days

6.2   The Problem with Relativity

For reasons of brevity, I shall refer to the theories of general and special relativity as 'Relativity' and the author of these theories as the 'Author'.

These Chapters (6.2.1 to 6.2.6) include numerous reasons why Relativity can no longer be considered appropriate for orbital motion, the most significant of which are:

1) *All* orbits work perfectly without it (i.e. it is unnecessary)

2) It causes many aspects of Newton's laws of orbital motion to fail

3) It was developed simply to support item 4) below

4) The Author misunderstood the meaning of both $E=mc^2$ & light

Just one, or even two, of the following problems (6.2.1 to .6) could be chalked up to coincidence, but all six!

A major concern regarding Relativity is the lack of attention paid to matching units in its formulae. For example, it appears to include the formula: '$R_{ab} - \frac{1}{2}R.g_{ab} = T_{ab}$'; in which length is added to velocity squared which results in time. Even based upon Reimann mathematics, this doesn't make much sense.

It is important to remember:

Whilst it is possible to create a sub-theory to explain any distortion of reality you wish, why would you if there is no need?

When everything in the universe can be explained without the sub-theory, the sub-theory becomes redundant.

Relativity was driven by a desire to explain events that were either unknown or misunderstood. Now that we fully understand the theory behind *all* orbital systems and *light*, Relativity has become redundant, especially as it actually invalidates Newton's laws of orbital motion, that otherwise work perfectly, *in every respect*; irrespective of energy, speed and mass.

It seems clear to me, that Relativity must be declared *'dead in the water'*

6.2.1 Light Deflection

Light is apparently observed to deflect by an angle of 1.75 arc-seconds when passing at or close to the surface of our sun (Fig 45; a).

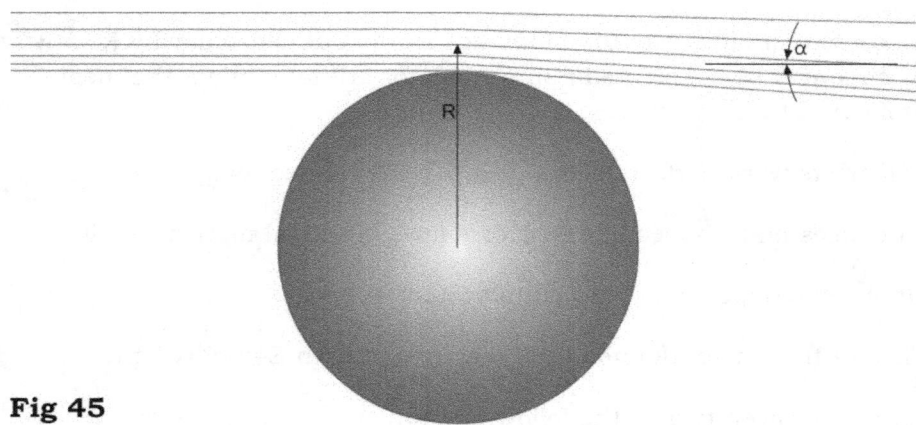

Fig 45

According to the Author, however, Isaac Newton's laws of gravity predict a deflection angle of half this value (0.875)

Relativity is a mathematical description of space-time/gravity distortion around celestial bodies that was developed to support the Author's formula for *observed* light deflection angles, which was based upon *'light-emitting photons'* and their susceptibility to gravity.

The problem with the Author's approach is that the light we see all around us is electro-magnetic energy, and therefore possesses no mass, so gravitational laws don't apply (refer to Chapter 6.2.2).

On the other hand, Isaac Newton's gravitational constant (G), which is based upon the properties of Quanta, may be used to define the same deflection angle (a) as follows:

$a = Atan(4.a_o.V_s / R)$ {m³.m / m³.m}   #

Where:

$G = a_o.c^2/\rho_u$ {m³ / kg.s² per m³}

$V_s = m_s/\rho_u$ {m³}

m_s = the mass of our sun {kg}

ρ_u = ultimate density {kg/m³}

a_o = Bohr's radius {m}

c = the speed of light in a vacuum {m/s}

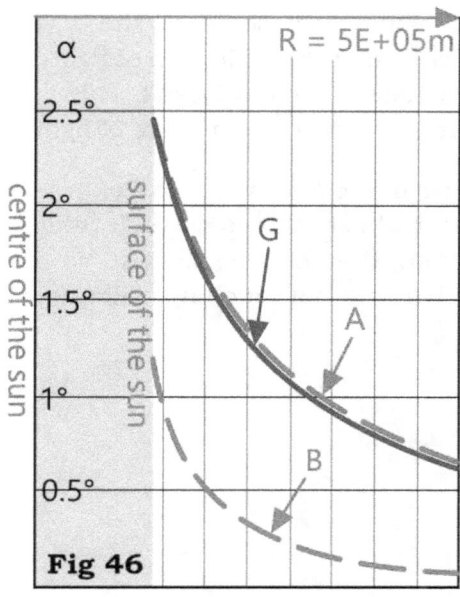

Fig 46

When calculating light deflection angles, the Author reduced the size of our sun by a factor of 5000, retaining its correct mass and increasing its density accordingly. Presumably, this was to raise the calculated angles to a practical value.

Fig 46: Curve **A** shows the variation in 'α' according to **Relativity**, from the surface of the above modified sun to a distance 5E+05m from its centre, which complies with the *observed* deflection value.

Fig 46: Curve **B** shows the equivalent variation in 'α', according to the Author, when using **Newton's laws of gravity** on a '*photon*'

Fig 46: Curve **G**; we can also reproduce the *observed* light deflection angles if we plot the light deflection angles using formula # (above), which is based upon **Newton's gravitational constant**. The difference being that this formula, which is not based upon the gravitational susceptibility of '*photons*', does not need a sub-theory to justify it.

It is important to understand that it isn't necessary to prove the validity of an alternative theory in order to discredit the original. You only need to demonstrate that the same result may be achieved by applying valid input data to an identical model but without the need for a sub-theory for justification. The above formula does exactly that.

That light travels in waves and not as particles is not new; Christiaan Huygens declared this to be the case in the late 17th century and it was later restated by Faraday, Maxwell and Pauling.

Newton on the other hand declared light to be particles. The Author automatically declared Newton to be correct because he quite rightly held Newton in such high esteem, and therefore used Newton's laws of gravity to deflect [particle] light as it passed celestial bodies. The problem was that this produced incorrect values (Fig 46: Curve **B**), so the Author created a sub-theory (Relativity) to justify his approach.

General relativity, the deformation of space-time, is based upon the inability to use Newton's laws of gravity to predict the deflection of light, which is only a problem if light possesses mass, which it doesn't. This theory is therefore based upon a misunderstanding of the nature of light.

General relativity was also devised because of the Author's disbelief in force-fields, which he called the 'ether'. But whilst there is no such thing as *the ether* (as he understood it) we do know that force fields exist, as anybody holding two magnets close together, but not touching, will know.

6.2.2　The *Speed of Light*

All physicists currently claim that the light we see is emitted by photons.

They also claim that electrons are 'weird beasts' that possess mass and travel in waves, which is the reason we cannot pin them down (uncertainty principle).

This is difficult to understand given that; if the entire electro-magnetic spectrum ranges between <2E-14m and >7m, how can all photons travel at the same velocity. Surely, they must travel at all speeds between >0 to 'c' in order to represent the full electro-magnetic spectrum.

For example; an electron travelling at 1E+06m/s will possess a different energy to one travelling at 1E+08m/s. The electro-magnetic energy (wavelength; e.g. colour) each radiates must, therefore, also be different. And if so, according to Newton's laws of gravitational attraction, different wavelengths of light must be deflected at different angles (refer to Chapter 6.2.1: α).
Which is contrary to the fundamental principle of both theories of Relativity: That "light possesses mass" and all wavelengths of light passing the same celestial body at the same radial distance is deflected at the same angle (α).

Therefore, the light we see cannot be photons, it must be electro-magnetic energy, which possesses no mass and is deflected by magnetic charge.

Not only is it unnecessary to deform space-time around celestial bodies in order to explain light deflection, it is mathematically incorrect to do so.

Special relativity was devised because of the inability to correlate the additive nature of mass-velocity with the non-additive nature of light. This is only a dilemma if light possesses mass, which it doesn't. The *photon* is therefore, also based upon a misunderstanding of the nature of light.

There is no such thing as a photon (refer to Chapter 6.4).

6.2.3   Neutronic Radius (Rn)

The neutronic radius (refer to Chapter 3.5.1.3), which is achieved by an orbiting electron when travelling at 'c', can *only* be explained using Newton's laws of orbital motion and Coulomb's law of electrical force. It occurs in far too many constants (magnetic, permittivity, Rydberg's, Planck's, Coulomb's, Henry's, etc.) to be rejected as a *fundamental physical constant*.

The neutronic radius is also the basis of $E=mc^2$ (refer to Chapter 6.2.5)

The conversion of mass to energy with velocity together with the space-time/gravitational distortion around force-centres as defined in Relativity, would render such an orbital radius impossible. I.e. the electron would be orbiting inside the proton at 'c' and R_n would be incorrect, making magnetic constant, permittivity, Rydberg, Planck, Coulomb, Henry, etc. incorrect, which we know is not the case.

6.2.4   Elliptical Orbits

Relativity is based upon the predication that light possesses mass (refer to Chapter 1.1.1) and that gravity is responsible for its deflection, and because the Author claimed that Newton's laws of gravitational attraction cannot apply to light (refer to Chapter 6.2.1), it was necessary to deform space-time around celestial bodies and artificially modify satellite velocity to account for this problem. But if we apply the relativistic velocity modification to a satellite, such as the earth (Fig 47); $v = v / \sqrt{[1+(v/c)^2]}$ we find that; whilst Newton's laws <u>always</u> work (centrifugal and gravitational acceleration always balance) irrespective of satellite velocity, Relativity fails above <1% of *light-speed*. In fact, according to Relativity an electron can never actually achieve *light-speed*: $c \neq c/\sqrt{2}$

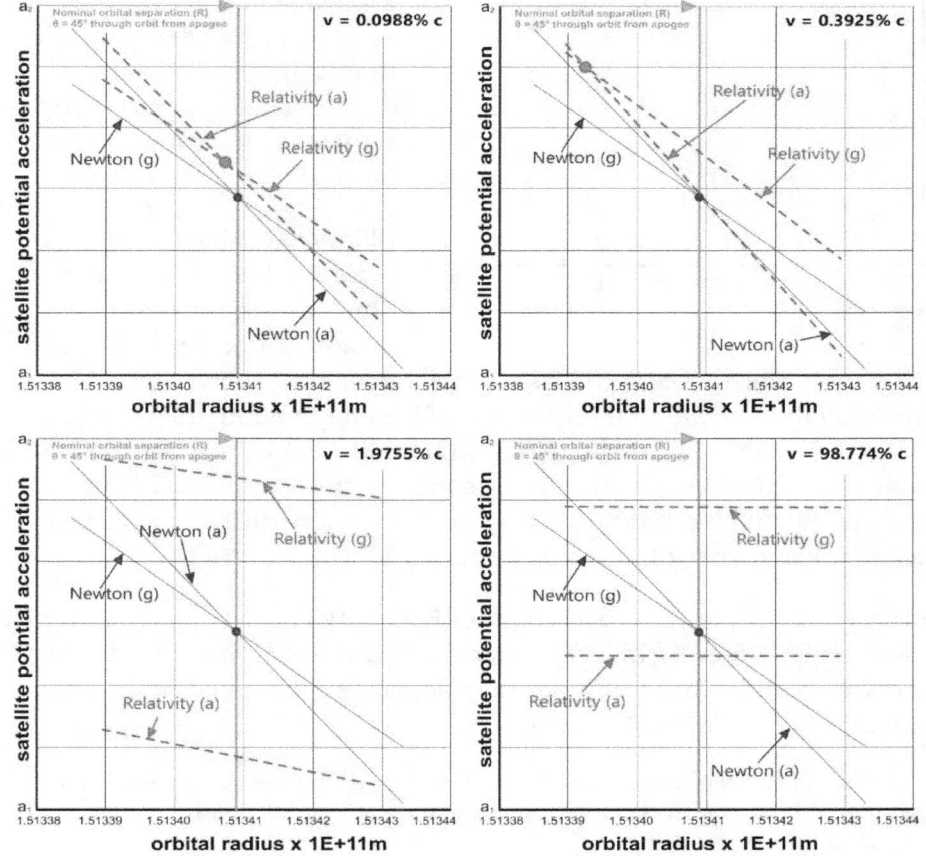

Fig 47

Fig 48 shows the same calculation performed for an electron passing the sun at the *speed of light* at an orbital radius of 765061000m (refer to Chapter 6.2.1; Fig 45) in which Newton's laws of orbital motion function correctly with no modification to elliptical orbits or satellite velocity. Relativism, however, shows gravitational acceleration at *light-speed* is <u>always</u> greater (more than twice) centrifugal acceleration, meaning that a *photon* cannot pass the sun without being absorbed by it; if light possess mass and gravity is responsible for its deflection (as the Author claimed).

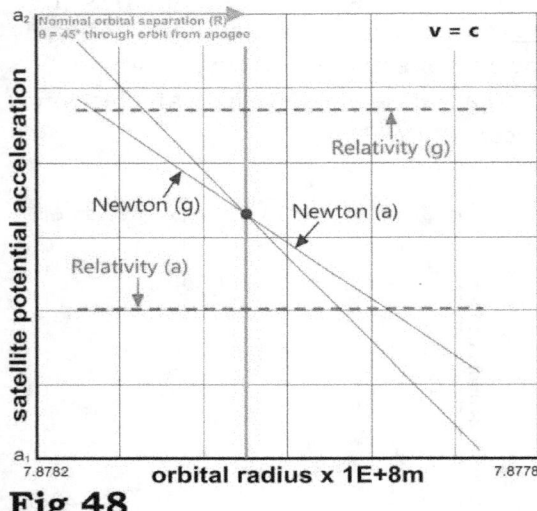

Fig 48

In fact, simply because the Author assumed that light possesses mass, he found it necessary not only to modify space and time but also artificially modify velocity, and even then, his theory fails in that it predicts an electron cannot achieve *light-speed*, destroying his concept of a *photon* (Figs 47 & 48).

Elliptical orbits are an indisputable fact of nature. This has been repeatedly demonstrated since Kepler's discovery in the 17th century. Its mathematical laws show that an exact ellipse is *fundamental* to the constant of [orbital] motion and thereby essential to maintain satellite paths in non-circular orbits. Relativity requires a distortion of this ellipse, rendering the orbital laws unworkable; yet we know that Newton's universal orbits work perfectly irrespective of size, shape and speed (refer to Chapter 4).

Moreover, if we alter time according to Relativity ($T_{AB} = R_{AB} - \frac{1}{2}.g_{AB}.R^{\#}$) the balance will shift ensuring that external interaction (from other bodies) cannot be rectified naturally; yet they do, as demonstrated in Chapter 2.2.8 and Figs 47 & 48 (Newton)

The units within this formula, which do not match, cannot be reconciled without a sub-theory. Such an anomaly does not occur in any of Isaac Newton's theories.

It must therefore be concluded that relativity is not only unnecessary; this aspect of the theory is also incorrect.

6.2.5   E=mc²

$E=mc^2$ was first postulated by Henri Poincaré towards the end of the 19th century ($c = \sqrt{[E/m]}$), however, he did not explain its physical relevance other than it represented a terminal velocity.

Numerous formulas were contrived in an attempt to rationalise this belief;
$m = m_o / \sqrt{[1\pm(v/c)^2]}$ & $v = 1 / \sqrt{[1\pm(v/c)^2]}$
but it appears to show that mass is infinite when; $v = c$ or v can never exceed '$c/\sqrt{2}$'

During the creation of this publication, I discovered that this relationship is the *potential* energy between a satellite and its force-centre in circular orbits (e.g. atoms), where; PE = -2.KE (refer to Chapter 2.2.3)
$KE = \frac{1}{2}.m.v^2$; $PE = m.v^2$
and at the speed of light, an electron orbits its proton at radius R_n, which is a fundamental constant (refer to Chapter 6.2.3) where $PE = m.c^2$ at the creation of a neutron.

Whilst Poincaré's formula ($E=m.c^2$) does indeed represent a terminal velocity, it refers to the ultimate potential energy between a proton and its electron that is *orbiting* at 'c'. And occurs when the magnetic attractive field energy exceeds centrifugal repulsion energy and the electron combines with its proton to create a neutron. It does not refer to an electron in free-flight travelling at the speed of light, or in fact, kinetic energy of any kind.

That said, electro-magnetic energy can only be radiated whilst an orbiting electron is travelling at less than 'c', which means that no electron can *naturally* achieve this kinetic energy (which it gets from electro-magnetic radiation): $E = \frac{1}{2}m.c^2$. However, this limitation *does not* mean that an electron, or anything else, cannot travel faster than 'c' if given sufficient energy artificially.

This means that in Relativity, $E=mc^2$ has been inappropriately applied to kinetic energy to describe mass-energy variation with velocity, which does not actually occur; mass does not vary with velocity and there is no such thing as mass.

Combining the theories from Newton, Planck and Poincaré:
Assuming 'm' is a unit of mass of ultimate density: $m = \rho_u$
Newton: $G = a_o.c^2 / m$
Planck: $F = c^4/G$
$F = m.c^4 / a_o.c^2$
$F = E/R$
$E = F.a_o$
$E = m.c^2$ (Poincaré)

The following is an hypothetical argument because black-holes are not sized to trap electrons (refer to Chapter 2.7.2), but the mathematical argument is valid because it describes the relationship between $E=mc^2$ and Schwarzschild's radius.

Kristian Huygens gave us the relationship between acceleration and velocity;
$v^2 = 2.a.R$ (for the PE component)
and Henri Poincaré showed us that $E = m.c^2$, which today is generally believed to represent kinetic relativism but we now know that potential energy is twice kinetic energy in circular orbits (e.g. atoms):
$PE = 2.KE = m.v^2$
If we assume a limiting gravitational energy that will trap an electron travelling at the speed of light, it is probably equivalent to that defined by Henri Poincaré, i.e. for any specified mass; $m.c^2 = m.2.g.R$
where 'g' is the gravitational acceleration at its outer surface (at radius 'R')

Therefore, for any specific gravitational energy, according to: $E = m.c^2$
we should be able to find the associated limiting mass with respect to its ability to emit light:
$c^2 = 2.g/R \rightarrow g = c^2 / 2.R$
$E = m.g.R \rightarrow m.R.c^2 / 2.R \rightarrow \frac{1}{2}.m.c^2$ (kinetic energy at light speed)
i.e. if $2.g.R \geq c^2$ for a given force centre, electrons will have insufficient energy to escape its surface.
if $g = G.m/R^2$ then $2.G.m/R = c^2$ represents the limiting mass

c^2 in this famous equation therefore represents a limiting gravitational acceleration that may be used to define the potential energy required to trap an electron travelling at the speed of light, and the formula becomes:
$E = m.g.R$
where $g.R = c^2$ and the term 'm.g' refers to the potential force on electrons

'E' in this formula is not kinetic energy, it is potential, i.e. Henri Poincaré's famous formula wasn't showing us what has euphemistically become relativism;
between them, Isaac Newton and Henri Poincaré were showing us how to size a [fictitious] black hole, because Newton incorrectly believed that light possessed mass!

$E = m.c^2 = m.2.R.g$

$c^2 = 2.R.g$

$g = G.m/R^2$

$c^2 = 2.R.G.m/R^2 = 2.G.m/R$

$R = 2.G.m/c^2$

Schwarzschild radius of a proton (Rs)

If 'm_p' is the mass of a proton:

$Rs = 2.G.m_p/c^2 = 2 \times 6.67359232E-11 \times 1.67262164E-27 \div 299792459^2$

$Rs = 2.48396784934951E-54$ m

When an electron is orbiting at the speed of light (c), the electron-proton separation radius '$R_n = 2.817729E-15$ m'
Moreover; $Rs = 2.\varphi.R_n$

This is the point at which an electron comes together with a proton to produce a neutron. The temperature of the electron orbiting at the speed of light is: $\underline{T} = X.c^2 = 620963351.43177$ K

All of which demonstrates that 'E=mc^2' applies to potential, not kinetic energy. Therefore, electrons (and all matter) in free-flight are not limited to the speed of electro-magnetic energy (e.g. light).

$E=mc^2$ has nothing to do with kinetic energy and mass does not change into energy with increasing speed.
$E=mc^2$ (which was discovered/prophesied by Henri Poincaré) refers to circular orbits where potential energy (PE) is twice kinetic energy
$(PE = 2.\frac{1}{2}.m.v^2)$.
At the velocity of light (v = c), a proton and its electron combine to create a neutron (refer to Chapter 6.2.3).

'E=mc^2' refers to potential energy, not kinetic energy.

6.2.6 Hades

At the time Relativity was theorised, neither its Author or anyone else was aware of the exigency of force-centres in *every* orbital system or of [planetary] spin theory. The Author therefore misunderstood the effect of galactic population on orbital shapes (refer to Chapter 2.2); hence the misguided invention of dark matter (refer to Chapter 2.7.4).

Moreover, if the Author had known of Hades and the laws of station-keeping (refer to Chapter 2.2.8), he would have realised that the deformation of space-time could not work.

6.3   The Problem with Quantum Theory

For reasons of brevity, I shall refer to Quantum Theory as 'QT' and its author as the 'Author'.

There are numerous reasons why QT can no longer be considered appropriate for the description of atoms, the most significant of which are listed below.

1) Whilst QT cannot explain or describe the behaviour of atoms in terms of what we see, feel and hear in the universe, an atom according to Isaac Newton and Coulomb can do this.

2) QT requires a still-undiscovered unification theory to ensure compliance with Newton's laws of motion.

3) The non-orbital nature of QT electrons means the QT atom cannot generate or emit electro-magnetic energy.

4) QT required the invention of 'string-theory' along with numerous sub-atomic particles (e.g. quarks, fermions, bosons, gluons, etc.) in order to make the atom work, whereas Newton and Coulomb can make the entire universe work with just two particles; the electron and the proton.

5) QT relies on statistics for justification; statistics apply only to the consequences of the laws of nature, never the laws themselves.

6) As is demonstrated in Chapter 4.5, Newton's and Coulomb's theories can be applied to the atomic structure described in this book and therewith mathematically predict its properties.

7) QT's Author needed intimidation to force acceptance of his theories by the scientific community;
"if you aren't profoundly shocked by quantum physics, then you haven't understood it"

8) QT remains unproven after 100 years.

9) It was necessary to invent sub-theories (including the uncertainty principle) to explain why electron location cannot be predicted in QT. This approach is similar to that devised by religious communities for their gods; "I refuse to prove I exist says God, for proof denies faith and without faith I am nothing". It is an untenable position.

The single biggest problem with QT (item 3 above) is that its atom can only absorb energy; it has no way to emit it. However, item 6 above also proves that Newton's is the correct atomic model.

QT was driven by a desire to explain events that were either unknown or misunderstood. Now that we fully understand the theory behind all orbital systems and that light does not possess mass, QT has become redundant, especially as it does not obey Newton's laws of orbital motion.

Whilst it is possible to create a sub-theory to explain any distortion of reality you wish, why would you if there is no need?

When everything in the universe can be explained without a sub-theory, the sub-theory becomes redundant.

It seems clear that Quantum Theory must be declared '*dead in the water*', given that it fails to address the atom's single most important issue; the emission of electro-magnetic energy, whilst Newton's and Coulomb's laws together can explain all aspects of atomic structure and performance.

6.4   The Error

Relativity and quantum theory came about because of an error made prior to the 20th century. I believe that this error is responsible for having stalled scientific progress for a hundred years.

It is currently believed that electrons emit light (photons); **they don't**.

We have been taught this for a hundred years, forcing us to create fanciful theories to explain how mass moves in waves; **it doesn't**.

The photon exists in our minds because of a very simple mistake made a long time ago relating to Crooke's tube.

Crooke, and everyone since, believed that he had created a perfect vacuum by pumping out all the air from his tube; **he hadn't**! His tube contained a *measured* vacuum (refer to Chapter 6.4.1), it was not a *true* vacuum; there were millions of protons inside it. It is impossible to create a perfect vacuum on planet Earth, or anywhere on or in a celestial body where all matter resides.

When Crooke fired electrons from one end of his tube to the other, they appeared to emit light. So, he and everyone since believed that electrons must emit light. But the light you are seeing is not emitted by electrons, it is the electro-magnetic radiation emitted due to their interaction with protons in the tube. When a bar magnet is placed beside the tube, the light path deflects. What you see is not the bending of light (although magnetism does [slightly] bend light), the dramatic deflection you see is that of the *path* of the electrons; the light is emitted by interactions along this deflected path.

During his 'light-on-a-metal-plate' experiment Max Planck detected a feint but perceptible electric charge that '*pulsated*', confirming that it was induced by electro-magnetic energy. If it had been induced by a stream of electrons (photons) it would have been continuous as in a battery, confirming that light is electro-magnetic energy not electrons. The light emitted by stars and everything else in the universe is not brought to us by electrons: it is radiated electro-magnetic energy. Michael Faraday understood this and James Clerk Maxwell described it mathematically.

These are the main reasons why relativity and quantum theory remain unproven after 100 years; they both rely on light-emitting electrons.

6.4.1 Measured Vacuum

A *measured* vacuum defines the number of protons inside Crooke's Tube:
Crooke would have had to remove every proton from inside his tube to
claim that electrons fired within it were emitting light.

His tube was originally filled with air:
78% nitrogen, 21% oxygen & 1% argon.
Fig 49 shows the number of protons inside his tube at pressures between
7.5E-18 bar and 1E-05 bar.

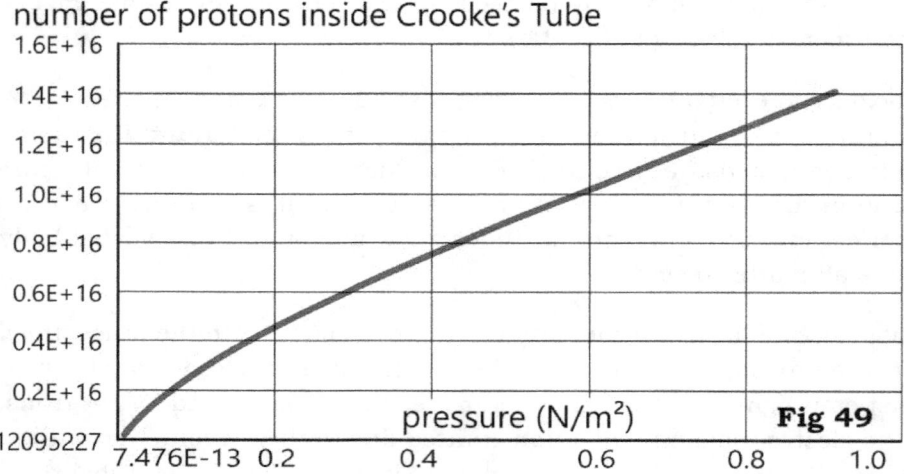

Gas pressure, including that in air, originates from the force generated
between adjacent atoms by their electrical-charge repulsion.

This repulsion (force) drops off very quickly with the [square of the]
distance between the atoms (exponential curve), and as can be seen from
the above graph, even at 7.5E-13 N/m² (7.5E-018 bar), there would still
have been about 12 million protons inside his tube.

Crooke used a mechanical pump to remove, what he thought was, all the
air from inside it.

However, today's best laboratory mechanical pumps could achieve little
better than 1E-10 bar (1E-05 N/mm²), which relates to about 3E+12
protons left inside his tube. It is expected that Crooke actually achieved
about 1E-05 bar, which meant there were about 1.4E+16 protons left
inside his tube.

6.5   Model Verification

The following sub-chapters provide supporting mathematical verification of the atomic model described in this publication. But the relationship provided below is the final conclusive proof that the atomic model described in this book is correct:

Dalton's law states that each gas in a mixture of different gases will evenly fill its container independently of all the other gases in the mixture.

Partial pressure theory states that the total pressure of a gas mixture is the sum of the pressures of each individual gas.

Coulomb's force law between the electrical charges developed in adjacent protons with temperature is responsible creating gas pressure.

P.V = n.R_i.T:

Today, we calculate the pressure (p) of a gas in a container thus:
$p = n.R_i.T / V$
where: 'n' is the number of moles in the gas, 'R_i' is the ideal gas constant, 'T' is its temperature and 'V' its volume.

But it can also be calculated thus:
$p = \rho.PE_1 / m_M.Y$
where: 'ρ' is the gas density, 'PE_1' is the potential energy between the proton and the electron in shell-1 and 'm_M' is the molecular mass
$PE = m_e.v_e^2$
$v_e = \sqrt{[T/X]}$
which provides *exactly* the same result as the PVRT calculation method but is much simpler because there is no need to play with moles.

Because this latter calculation method replaces 'PVRT' but bypasses Boltzmann's constant, Avogadro's number, gas temperature and the ideal gas constant with potential energy and electron velocity, the model described in this book must be correct.

However, there are a few more proofs available ...

6.5.1 Density vs Temperature

The magnetic field energy (MFE) generated by proton-electron pairs holds adjacent atoms together (viscosity) and is constant irrespective of temperature. The electrical charges (EC) generated in protons repel adjacent atoms (gases) and varies between e and e' with temperature.

Given that temperature effects (e.g. gasification) are dependent upon repulsive EC between adjacent atoms and density is dependent upon an attractive MFE between those same atoms, and that both charges are created by the same energy generation process (proton-electron pairs), density and temperature should follow similar patterns of behaviour according to the number of nucleic protons (atomic number (Z)).

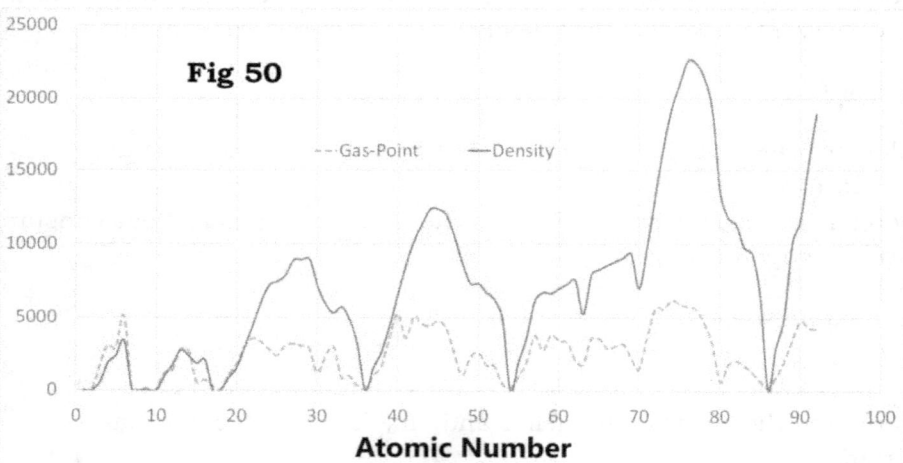

This relationship (between EC and MFE), which can clearly be seen in the temperature/density vs atomic number plot shown in Fig 50 is governed by the nucleic structure, which is the last significant piece of the atomic puzzle. Whilst it can be resolved mathematically (refer to Chapter 3.5.3), it has not been addressed here because it is not part of Isaac Newton's laws of orbital motion.

6.5.2 Specific Heat Capacity

The specific heat capacity of an atom defines the amount of energy it can hold in relation to its mass per unit temperature. This means the sum of the kinetic energy of all electrons in an atom's shells relative to its mass and *'temperature'*. The *'temperature'* in this case (as in all cases) is as defined in Chapter 2.1.6

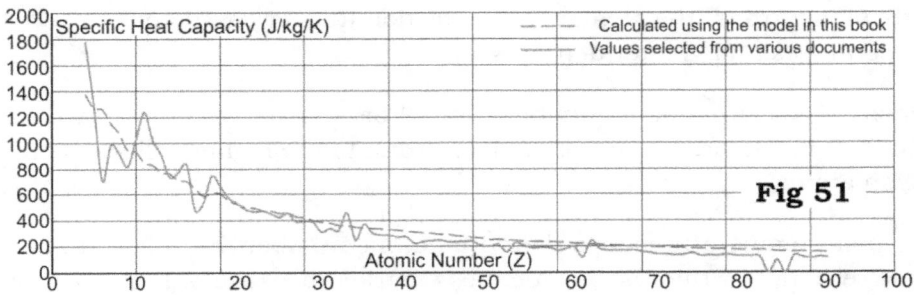

Fig 51 shows the calculated values for specific heat for all atoms from Z=4 to 92 compared to the documented values that have been taken from various sources and which are subject to experimental error.

This calculation technique, is as follows:

$$SHC = KE_T \, / \, Y.m.T_1 \quad \{J/kg/K\}$$

where:
KE_T = the total kinetic energy in every electron in the elemental shells
T_1 = the temperature of the electron(s) in the innermost shell
Y (refer to Chapter 5.4)
m = the total mass of the atom (including electrons and neutrons)

6.5.3 Gas-Point

The gas-point of any atom is the temperature at which its electrical charge (EC) exceeds its magnetic field energy (MFE).

If the MFE is greater than the total exposed EC, the atoms will exist as viscous matter; otherwise they will exist as a gas.

Outlying nucleic neutrons protect adjacent atoms from EC. The more outlying neutrons, the greater the protection (higher gas-point temperatures and greater densities).

Density rises with atomic number (Fig 50) because larger atoms tend to collect a greater percentage of neutrons, due to the higher collective MFE within atoms.

Gas-point temperatures also rise with atomic number (Fig 52) for the same reason, but this rise is much less marked because only the outlying and exposed proton EC actively repels.

A mathematical relationship that reflects reality has been identified for all the atoms in the Periodic Table (refer to Chapter 3.5.3).

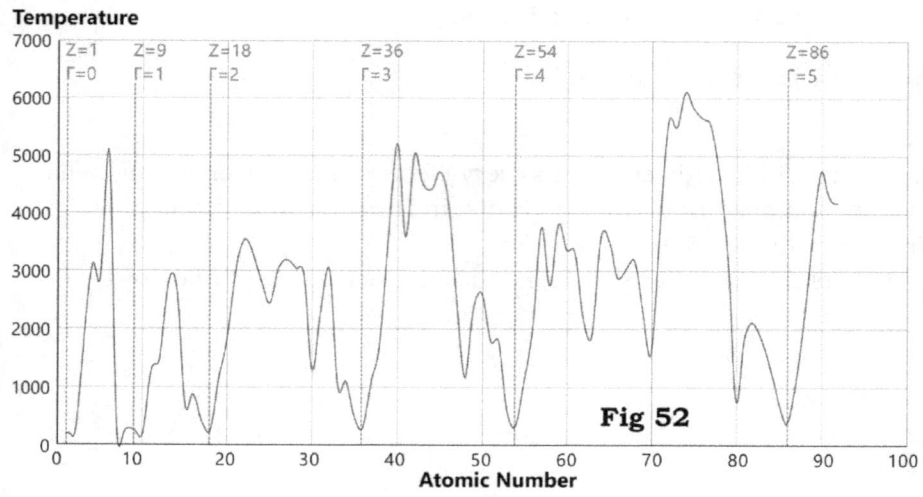

Fig 52

This relationship (Z → Γ) forms the basis for mathematical chemistry.

6.5.4   Our Sun

Almost all (>99.7%) natural hydrogen exists as lone protons (H⁺), which cannot be solidified or liquefied due to their similar positive charges. Lone protons also have no way of absorbing or emitting electro-magnetic energy (heat and light).

It is currently claimed that our sun was created from a cloud of hydrogen atoms that accreted into a star due to gravity and an *external force*. This is of course, impossible (refer to Chapter 2.7.5)

The surface temperature of our sun is said to be about 5778K, which would be impossible if the sun's surface comprised lone protons that cannot collect or emit electro-magnetic energy (i.e. heat or colour).

Yet according to the atomic model proposed in this book, at 5778K;
KE = 3.7493802154296E-19 J
electron velocity = 912757.252 m/s
at an orbital radius of 3.03992067E-10 m
f = v / 2πR = 4.77873747733E+14 Hz
λ = c/f = **6.2734657516211E-07 m**

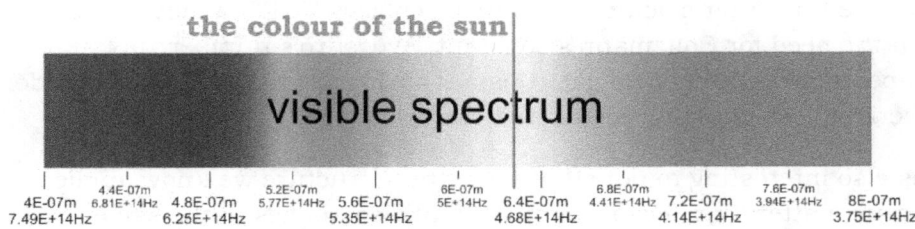

Fig 53

This demonstrates that the hydrogen at the surface of the sun is predominantly proton-electron pairs, which in turn means that these hydrogen atoms (proton-electron pairs) *must be a by-product of fission.*

6.5.5 PVRT

The final proof of this atomic model can be seen by the replacement of the well-known theory P.V = n.Ri.T with an alternative calculation using the potential energy in a proton-electron pair as described in Chapter 3.5.2

Today, we calculate the pressure (p) of a gas thus:
p = n.Ri.T / V
where: 'n' is the number of moles in the gas, 'Ri' is the ideal gas constant, 'T' is its temperature and 'V' its volume.

But it can also be calculated thus:
p = ρ.PE₁ / mM.Y = $T.m_e.\rho$ / $X.Y.m_M$ = $k_B.T.\rho$ / m_M

where: 'ρ' is the gas density, 'PE₁' is the potential energy between the proton and the electron in shell-1 and 'mM' is the molecular mass:
$PE = m_e.v_e^2$
$v_e = \sqrt{[T/X]}$

which provides *exactly* the same result as the PVRT calculation method but is much simpler because there is no need to play with moles.

Because this latter calculation method replaces 'PVRT' altogether, along with the need for Boltzmann's constant, Avogadro's number, gas temperature and the ideal gas constant with potential energy, the model described here must be considered correct.

It is also interesting to note that the lattice structure we know applies to viscous matter (ζ : refer to Chapter 3.5.3) also applies to the same elements in gaseous form, and is responsible for *partial pressure*.

6.6   Heat

Heat is the electro-magnetic energy radiated by a proton-electron pair: the greater the energy, the greater the heat

Temperature is what we feel from the electro-magnetic energy radiated by a proton-electron pair with its electron(s) in the innermost shell: the greater the energy, the greater the temperature

We sense this heat and temperature through the energy in the electro-magnetism radiated by the proton-electron pairs.

Heat energy in an atom is the sum of the kinetic energies in all of its electrons.

The relationship between electron velocity (kinetic energy) and temperature may be defined as; $T = X.v^2/e^2$ {K}
but as 'e' is a constant:
$T = X.v^2$ {K}
Which is the same as: $T = PE / k_B.^3\sqrt{[\ \frac{1}{2}\ .\ \sqrt{[(4\pi)^2.a_o\ /\ R_n]}\ ]}$
Where:
$X = T.t_n^2/(2\pi.R_n)^2 = T/c^2 = 6.9353271647894E\text{-}09$ K.s^2/m^2
and KE is the kinetic energy of the electron

The relationship between electron orbital radius and temperature may be defined as:
$T = X_R/R$ {K}
Where: $X_R = T_n.R_n = 1.75646616508036E\text{-}06$ {K.m}

You may have noticed that; $T = X.v^2 / e^2$ is similar to Newton's gravitational force $F = G.m_1.m_2 / R^2$, Coulomb's force $F = k.Q_1.Q_2 / R^2$, and Gilbert's and Maxwell's formulas for force and energy. It is therefore anticipated that all of these formulas will eventually become just two; one for magnetic charge (gravity) and the other for electrical charge.

An interesting relationship for the above heat constants is as follows:
$4\pi^2.X/X_R = K$ {s^2/m^3)

Where K is Isaac Newton's orbital constant of proportionality for circular orbits (e.g. as in the atom):
$K = t^2/a^3 = 0.15587874533403$ {s^2/m^3)

All heat energy is radiated. Convection and conduction are simply different forms of radiation.

Conduction is simply the transfer of electro-magnetic radiation between proton-electron pairs within matter.

Convection is simply the balancing of electrical field energy between adjacent atoms. Atoms at a higher temperature (with greater heat energy) will try to move to a position where other atoms are further apart (e.g. cooler locations). In our atmosphere, this is always *upwards*, away from the earth's surface where the electrical repulsion energy (in a gas) can balance with the magnetic attraction energy (gravity).

6.7   Mass

I claim that there is no such thing as mass,
i.e. **mass** is magnetic charge, the magnitude of which is equal to the
elementary charge unit (e):

$m = |e|$

$m' = |e'|$

Therefore; what is currently referred to as the elementary charge unit
should be the *electrical charge unit* (±e), and the unit of mass should be
the *magnetic charge unit* (m). The magnetic charge's non-polar nature is
what causes all particles to attract all other particles.

Every particle holds a constant non-polar magnetic charge and also
retains the capacity to hold the same magnitude of electrical charge.

For example;
The electron holds '*m*' Coulombs of magnetic charge and '-e' Coulomb's of
electrical charge constantly.
The proton holds '*m'*' Coulombs of magnetic charge and '+e' Coulomb's of
electrical charge. However, its greater magnetic charge, gives it the
capacity to increase its electrical charge to '+e'' if and when partnered by
an orbiting electron.

The number of particle assemblies (electron + proton + neutron) in a body:
n = mass / $2.(m_e+m_p)$

The magnetic charge in each particle assembly:
$m = 2.(e+e')$ = 5.88688075484235E-16 C

Using the planet Mercury to demonstrate this concept …

First; we generate a new magnetic constant (M) based upon Coulomb's
constant (k) as follows:

$M = c^2.m_e^2.R_n / m_p.e^2$ = 4.89477777726655E+06 kg.m³ / C².s²

Note:
If $R_K = \sqrt{[G/M]}$ = 3.69243914581423E-09 C/kg
{√[(m³ / s².kg)/ (kg.m³ / C².s²)] = C/kg}
and RC = 1.75881869180547E+11 C/kg
R_K / RC = 2.09938589066497E-20 = $\sqrt{\varphi}$
i.e. the square-root of the coupling ratio (refer to Chapter 6.11.3)

Next: we calculate the magnetic charge in Mercury and its force-centre:

	Mercury	Sun
mass (kg)	3.3011E+23	1.9885E+30
No. of particle groups	9.862670275E+49	5.941025671E+56
magnetic charge (C):	m_2 = 1.21891108642474E+15	m_1 = 7.34241524145157E+21

Table 6.7-1: Magnetic charge of the planet Mercury and its force-centre

Finally: we use our new constant 'M' and the magnetic charges (m_1 & m_2) to calculate the 'gravitational' force between Mercury and the sun:

$F = M.m_1.m_2/R^2$ (R is the separation distance at Mercury's perigee)

F = **2.07016816968015E+22** N
(refer to Chapter 4.2.2; F^P = **-2.0701682E+22** N)
and it works for all the planets in our solar system

There is an additional argument:

Potential Energy:
The potential energy of every orbiting body may be calculated using the following formula:
$PE = [(m_1.m_2) \div (m_1+m_2)] . g.R$ {J}   **#1**
Where; g is the gravitational acceleration from m_1 at radius R
This formula (**#1**) produces exactly the same result as Isaac Newton's formula for the same property:
$PE = G.m_1.m_2/R$ {J}

$PE = \sqrt{[(E_1.E_2) \div (E_1+E_2)] . g.R/v_C^2}$ {J}   **#2**
E_1, E_2 & v_C : refer to Chapter 6.11.15

Both formulas (**#1** & **#2**) produce exactly the same potential energy between every force-centre and every satellite in our solar system confirming that mass may also be defined as magnetic energy.

Moreover, if 'n' represents the number of atomic particles in a celestial body and e is their Coulomb value; for every planet in the solar system:

$g.R.e.(n_1.n_2)/(n_1+n_2) = 2.08908021007445E-08$ {$C.m^2/s^2$}

6.8   Gravity

Gravity is the potential energy in all matter due to the non-polar magnetism in atomic particles.

Non-Polar Magnetism is universal; it is always present in every atomic particle and it is all-pervasive; i.e. it radiates out in all directions and its strength does not diminish with distance. Whilst this form of magnetism is weak, it accrues collectively. I.e. the magnetic strength of 100 particles is 100 times stronger than that in one particle. This is why planets and stars - that comprise enormous numbers of particles - have such a strong magnetic attraction, but is not the case in, say, a cup.

Contrary to popular belief, this [non-polar] magnetic energy does not decrease with the square of the distance from its source. It only *appears* to do so because at increasingly larger radii the same force is distributed over a larger spherical area

Its effect is ever-present and constant throughout the universe, gradually pulling together all its matter, i.e. slowing down the expansion caused by the '*Big Bang*'. Eventually, all the particles in the universe will stop moving away from each other and gradually re-aggregate into another ultimate-body. The universe will then start all over again.

The concept *gravity is magnetism* changes nothing in terms of force and energy in the laws of orbital motion. Isaac Newton was correct. I.e. *exactly* the same results may be achieved either by using gravity (as Isaac Newton described it) or using magnetism (as William Gilbert and Hendrik Lorentz described it). The difference between gravity and magnetism is that we know exactly what magnetism is. We understand it and we can explain it in terms of energy. We don't, however, know what gravity is. Nobody, not even the great man himself, could explain what it is, what generates it and how it works.

All mass comprises a number of proton-electron pairs (neutrons are protons and electrons combined), each of which carries an elementary electrical charge (e). This electrical charge is responsible for both the electrical and the magnetic energy held by atomic particles.

Before explaining how the magnetic field is established, you need to know a couple of things:

'R_n' is the orbital radius of an electron travelling at the speed of light, when it combines with its proton to create a neutron. I.e. when the attractive magnetic energy exceed the repulsive electrical energy:
$$R_n = RC.\mu_o.e \ / \ 4\pi = G.m_p \ / \ c^2.\varphi = 2.817937953839E\text{-}15 \ m$$

I.e. the magnetic constant defines the radius where magnetic attraction between an electron orbiting at the speed of light equals electrical repulsion.

Magnetic Field:
$B = 1/RC$ {kg/C}

Faraday's constant:
$F = N_A.e = 96485.3317942156$ {C/mol}

Relative atomic mass of an electron:
$RAM_e = RAM_H.m_e/m_p = 0.000548580318390698$ {g/mol}
relative atomic mass of a hydrogen atom:
$RAM_H = 1.00727638277233E\text{-}03$ {kg/mol}

Avogadro's number (constant):
$N_A = 6.02214129E\text{+}23$ {/mol}

Elementary charge:
$e = 1.60217648753E\text{-}19$ {C}

Electron mass: $m_e = 9.1093897E\text{-}31$ {kg}
Proton mass: $m_p = 1.67262163783E\text{-}27$ {kg}

Using the above information, it is possible to explain gravitational energy in terms of *non-polar* magnetic attraction. Lorentz gave us the following formula for magnetic attractive force:
$F = q.v.B$
Where: 'q' is the electrical charge, 'v' is the relative velocity and 'B' is the magnetic field. However, in our case, because the attracting masses are stationary, the relative velocity in this formula must be modified to use the gravitational acceleration between the two masses.

It can get a little tricky from here …

Let me explain:

The magnetism referred to below is <u>non-polar</u> and all-pervasive. A polar [field] magnet is one that has all of its atomic nuclei aligned such that electro-magnetic energy is generated in one direction (e.g. north to south). Non-polar magnetism (charge) is always present in all atomic particles and acts in all directions. It is what holds the universe together, but it is much weaker than polar magnetism the amount by which it is weaker is called a coupling ratio: $\varphi = 4.407421117923350E\text{-}40$

Every mass (m) contains a specific number of particles (protons, electrons and neutrons). Because we already know that every particle possesses an electrical charge and that a neutron is a proton plus an electron, we can determine the number of electrical charges in any mass thus:
$q = m / (m_p + m_e)$
where m_p & m_e are the proton and electron masses respectively

Ok, here we go ...
Lorentz's magnetic force formula is;
$F = q.v.B$
where: q is the electrical charge; v is the velocity of the particle, and; B is the magnetic field.
His constant for the *magnetic field*:
$B = \mu_o.I / 2.\pi.R = 1/RC = 5.685634367312E\text{-}12 \ \{kg/C\}$
where: $R = 2.R_n$ & $I = e$
However, this approach is not very helpful for calculating the magnetic force between stationary particles.
Note: B is currently referred to as the magnetic field. Whereas, it is actually the reciprocal of relative charge $(B = 1/RC \ \{kg/C\})$

So we should use the following formulas for potential (rather than kinetic) force and energy:
$F = e.g/RC \ (kg.m/s^2)$
$E = e.g.R/RC \ (kg.m^2/s^2)$

On the other hand, the following approach can be corroborated with Newton's and Coulomb's formulas for the magnetic potential energy between a proton and its orbiting electron:
$PE = e.v^2/RC \ (= R/RC) \ \{J\}$
&
$F = PE/R$

Where: v is the velocity of the orbiting electron; RC is the relative atomic charge (B = 1/RC); g is the gravitational acceleration between the proton and the electron, and; R is the electron's orbital radius.
This is simply Lorentz's formula written in a useable format.

Because *exactly* the same magnetic forces and energies can be obtained universally for Newtons' gravitational calculations (in both astronomic and atomic environments), it is clear that gravity is actually magnetism.

Any two attracting masses (m_1 & m_2) may be described in terms of their static electrical charges (q_1 & q_2) which, in the case of an atom are equal to the elementary charge unit (e).

Exactly the same potential energy (PE) between m_p & m_e can be found using any and all of the following formulas:

Orbital: PE = $m_p.g$ (Isaac Newton)

Gravity: PE = G/φ . $m_p.m_e/R$ (Isaac Newton)

Electrical: PE = $k.q_p.q_e/R$ (Coulomb)

Magnetic: PE = q_p/RC . $g.R$ (Lorentz & Keith Dixon-Roche)

Heat: PE = $\underline{T}.k_B.m_p/m_e$ (Keith Dixon-Roche)

Where:
g is the gravitational acceleration between m_1 and m_2
R is radial separation between m_1 and m_2
B is the magnetic field (refer to Chapter 5.3: 1/RC)
k is Coulomb's constant
φ is the coupling ratio
m_p is the mass of a proton
m_e is the mass of an electron
q_p is the charge of a proton
q_e is the charge of an electron

The two important constants here are Isaac Newton's gravitational constant (G) and Coulomb's constant (k), both of which are based upon the properties of Quanta. It is important, therefore, to establish their relationship to each other.

$G = a_o.c^2 / \rho_u = 6.67359232004332E\text{-}11$ {m^3 / $kg.s^2$ per m^3}

$k = R_n.m_e.c^2/e^2 = 8.98755184732667E\text{+}09$ {$kg.m^3$ / $s^2.C^2$}
(refer to Chapter 6.11.4 and above)

so we can define the relationship as follows:
If: $Z . a_o.c^2 / \rho_u = m_e.R_n.c^2/e^2$
Then: $Z = m_e.R_n.c^2.\rho_u / a_o.c^2.e^2$
{$kg.\text{m}.\text{m}^2.kg.\text{m}^3.\text{s}^2 / \text{m}^3.\text{m}.\text{m}^2.\text{s}^2.C^2 = kg^2/C^2$}

Newton's gravitational constant is also equal to:
$G = k.e^2.\varphi / m_e.m_p = 6.67359232004332E\text{-}11$ m^3 / $kg.s^2$
which applies to his formula: $F = G.m_1.m_2 / R^2$

After dividing out the mass components:
$m_1.m_2 \div m_e.m_p$, we get a product of particles '$n_1.n_2$'.
If G_e now $= k.e^2.\varphi = 1.01682605280249E\text{-}67$ $kg.m^3$ / s^2
And rewriting Newton's formula thus: $F = G_e.n_1.n_2 / R^2$
We get the same result, but it is now in terms of a number of elementary charge units

This calculation for the electrical potential energy between the sun and the earth at its perigee, gives: PE = 1.2208949335E+73 J, whereas the gravitational potential energy is: PE = 5.380981972219E+33 J, the difference between the two being 'φ'

Whilst magnetic energy is accrued, electrical energy is shared, locking the electrical attractive energy between an electron and its proton at the atomic level. This [electrical] energy does not pass beyond the atom. I.e. 'G_e' may only be used instead of 'G' in Isaac newton's formula for the atom. It cannot be used for the calculation of lunar, solar or galactic orbital systems.

The above conclusively demonstrates that Isaac Newton's laws of orbital motion apply equally well to atoms

It is no longer necessary to use the term gravity; Gravity is Magnetism.

Moreover, it conclusively unites Newton's gravitational laws of orbital motion with those for the atom, something that cannot, and no doubt never will be the case for quantum theory.

PHILOSOPHIÆ NATURALIS PRINCIPIA MATHEMATICA Revision IV

6.9　Electricity

What is electricity?

Atomic particles possess static, constant magnetic and electrical charges.

The movement of a static electrical charge relative to a static magnetic charge will generate magnetic field lines around the electrical charge. This phenomenon also works the other way around. Electrical field lines will similarly encircle a moving magnetic charge. These field lines follow the right-hand rule and we use this phenomenon to generate field electricity (AC).

If sufficient voltage is applied to both ends of an electrical conductor, dependent upon the number of coulombs in the conductor, it is possible to overcome the potential energy between the electrons in the conductor and their protons. When this happens, the electrons, being negatively charged will be attracted to the next available atomic shell valency in the positive direction of the conductor.

Voltage is generally written as PD (potential difference). Potential energy is the same thing, as can be seen from its energy-related units; J/C

By applying a potential energy across a conductor, you are attempting to overcome the PE in the proton-electron pairs in its atoms. With sufficient energy (per electron or Coulomb), you may be able to cause more than one or two electrons per atom to swap atomic orbits.

The ability to overcome atomic PE is also dependent upon the temperature of the conductor. As its temperature rises, PE between electrons and their protons increases, making it necessary to raise the energy (voltage) to overcome the proton-electron PE. This phenomenon is called electrical resistance.

The units for electricity (V, A, Ω, H, F, etc.) are useless when trying to calculate electrical energy. I have therefore described below all of these properties in terms of units we can use in tandem with mechanical and energy systems.

Given our new-found knowledge of the atom, we can test this theory:

1 Amp = 1 C/s
and
1 Volt is 1 J/C
and
1 Watt is 1 J/s
and
1 Ohm is 1 J.s/C²
and we know that each electron has a constant static charge of:
e = 1.60217648753E-19 C
We can use an example to demonstrate how it works.

A calculation for copper wire and a tungsten filament using Newton's and Coulomb's laws of orbital motion and are given below:

	Formula	Copper @ 273K	Tungsten @ 3000K	
Dia. (Ø)		0.002	0.000046	m
Length (L)		5	0.533	m
Density (ρ)		8960	19293	kg/m³
at mass	ρ.L.Π.Ø²/4	0.140743351	1.70896244E-05	kg
PE_n		-2.391841E-21	-1.064981E-20	J
n	m / 2.m_n.Z	1.7606391E+20	6.8998021E+19	at
PE	n.PR_n	-2.391841E-21	-114.23350	J
Power (W)		500		J/s
Volts (V)		250		J/C
Amps (A)	P/V	2		C/s
Flow-rate	A/e	1.24830193E+19		el/s
per atom	Flow/n	0.070900498	0.180918513	el/s/at
at = atom		m_n = mass of a neutron		
el = electron		PE_n = potential energy in outer shell		

The filament will consume 2 amps and the electron flow-rate will be 1 electron per atom every 5.5 seconds. Electrons flow from each atom in the copper wire every 14.1 seconds.

When an electron is pulled from its orbit, the potential energy between it and its proton (PE_n) is released. This is how electrical heat (electro-magnetic energy) is generated (lost to the atom).

If the conductor temperature increases but the electrical power remains the same, PE_n will rise and the voltage necessary to transfer electrons will also rise, but the current that passes will fall (electrical resistance).

Because everything in the universe is electro-magnetic energy, our insistence on understanding everything in terms of units that currently do not apply (V, A, Ω, H, F, etc.) to electricity makes very little sense. I have therefore decided to describe the principal properties of electricity in the same units as everything else in this publication (refer to Chapter 5.8) Where:

Q = electro-magnetic radiation charge
f = electro-magnetic radiation frequency
E = electro-magnetic radiation energy
N_A = Avogadro's number
μ_o = magnetic constant (refer to Chapter 6.11.4)
R_n = neutronic radius (refer to Chapter 6.11.10)
m_e = mass of an electron
e = elementary charge unit

Current is the measure of electron flow rate along a conductor. So, this conversion is simple:
Amps: $\mathbf{A} = e.f$ {**C/s**}

Voltage is the potential energy needed to generate current. This energy is shared between all electrons in the conductor and will only cause electrons to pull out of their proton-electron pair orbits (generate electricity) within the conductor if the energy is high enough. So, this conversion is also simple:
Volts: $\mathbf{V} = E/e$ {**J/C**}

Temperature: $T = 2.E/k_B$ {J / J/K = K}
Where: E = electron kinetic energy

We know that **Resistance** is voltage divided by current:
$\Omega = V/A$ {J/C / C/s = **J.s/C²**}
But it can also be calculated in a few other ways:
$\Omega = \frac{1}{2}.m_e.(v/e')^2 / f = \pi.k'/v = E / f.e'^2$
Where: v = electron velocity and E is electron kinetic energy

Power is Volts x Amps: P = C/s . J/C = J/s (Watts)

The **Farad** relates to the electrical charge capacity of matter, and we know that **electrostatic capacitance** is the electrical charge in an electron multiplied by the mole, so this conversion is also simple:
Farad: $\mathbf{F} = e.N_A$ {**C/mol**}

It is also interesting to note that the Farad is the relative charge capacity of the electron, which can be demonstrated as follows:

RAC_e = 1F = N_A.e = 6.02214129E+23 x 1.60217648753E-19
 = 96485.3317942156 {C/mol}
This unit is the same as the electrical equivalent of RAM; *Relative Atomic Charge*, as it applies to the electron.

The RAC value for the proton is:
RAC_p = N_A.e . m_p/m_e = N_A.e' = 177161652.983418 {C/mol}

The specific gas constant for the electron: R_a = R_i/RAC_e {J/C/K}
The specific gas constant for the proton: R_a = R_i/RAC_p {J/C/K}

The **Henry** relates to the electrical resistance of matter, and we know **mutual induction** (Henry) as the rate of current change induced in an electrical circuit, so the units must be C/s^2.

Given the units for magnetic constant; H/m:
μ_o = 4.π . 1E-07 H/m
1E-07 = m_e.R_n/e^2 kg.m/C^2
μ_o = 4.π.R_n.m_e/e^2 kg.m/C^2
Henry: **H** = 1/μ_o **{kg.m^2/C^2}**

1 Henry is 1 Volt per Amp per second or 1 Coulomb per second squared (V.t / A or C/s^2)
The Volt per Amp is electrical resistance (Ω = V/A)
Their units are: V {J/C} & A {C/s}
V/A {J.s/C^2}
H = V.t / A {J.s^2/C^2 = kg.m^2/C^2}
where 't' is the period of rate of change of current (as in acceleration)

6.10   Newton vs Planck

Newton's atom is the real one; the one we see, hear and feel all around us. Planck's atom is theoretical, it does not exist. It is based upon his three constants (Table 6.10.2 below); time (t^P), length (λ^P) and mass (m^P).

It should be noted here that Planck's length is actually an orbital radius (R) and not a wavelength (λ) as depicted below. The reason being; force (F) is Energy divided by length, and a length in such a formula can only be an orbital radius.

The atomic particles in Planck's atom are identical in size; his electron is exactly the same, in all respects, as his proton. The only consistency between the two atoms (Newton's and Planck's) is that the product of the particle volumes in each atom is identical; $V_e.V_p = V^P.V^P = 3.0E\text{-}91 \ m^6$ (Refer to Chapter 6.11.1)

It is the comparison between these two atoms that has given us the ability to solve Newton's gravitational constant (G; refer to Chapter 6.11.2)

The following Tables (6.10A to 6.10C) provide comparative results for three atomic variations in which the particle properties are defined thus:

Newton atom; comprising Newtonian particles (Table 6.10A)
$t = a_o/c$; $\lambda = a_o$; $m = \rho_u$; $E = \rho_u.c^2$; $F = E/a_o$

Planck^N atom; comprising Newtonian particles (Table 6.10A)
t^P; λ^P; m^P; E^P; F^P (Table 6.10B)

Planck^P atom; comprising Planck particles (Table 6.10C), and calculated using a revised version of his own constant; ħ:
$ħ = \sqrt{[\ \pi.m.e^2.a_o\ /\ \varepsilon_o\ ]} = 7.99473592559180E\text{-}16 \ kg.m^2/s$
in which:
$a_o = \lambda^P\ /\ 4\pi^2$
$e = \sqrt{[\ G.m^{P2}\ /\ k.\varphi\ ]}$
$m = m^P$

Note: 'c', 'ε_o', 'G', 'φ' and 'k' are equal in both Newton and Planck atoms

Newton atom compared with a Planck^N atom:

	Newton Atom (A)	Planck^N Atom (B)	A/B
t	1.76514516887831E-19	5.39096122598359E-44	3.27426797353056E+24
λ	5.2917721067E-11	1.61616952231128E-35	3.27426797353056E+24
m	7.1266079635045E+16	2.1765500017459E-08	3.27426797353056E+24
E	6.40507585675678E+33	1.95618559889902E+09	3.27426797353056E+24
F	1.21038391820525E+44	1.21038391820525E+44	1

Table 6.10A

Newton atom compared with a Planck^P atom:

	Newton Atom (A)	Planck^P Atom (B)	A/B
t	1.76514516887831E-19	1.48432887846076E-34	1.18918737922067E+15
λ	5.2917721067E-11	4.44990604438464E-26	1.18918737922067E+15
m	7.1266079635045E+16	59.9283854507006	1.18918737922067E+15
E	6.40507585675678E+33	5.38609471364748E+18	1.18918737922067E+15
F	1.21038391820525E+44	1.21038391820525E+44	1

Table 6.10B

Planck^N atom compared with a Planck^P atom:

	Planck^N Atom (A)	Planck^P Atom (B)	A/B
t	5.39096122598359E-44	1.48432887846076E-34	2.75336589568949E+09
λ	1.61616952231128E-35	4.44990604438464E-26	2.75336589568949E+09
m	2.1765500017459E-08	59.9283854507006	2.75336589568949E+09
E	1.95618559889902E+09	5.38609471364748E+18	2.75336589568949E+09
F	1.21038391820525E+44	1.21038391820525E+44	1

Table 6.10C

Note: 3.27426797353056E+24 ÷ 1.18918737922067E+15 = 2.75336589568949E+09

6.10.1 Newton's Atom (particle properties)

The following Table provides the properties of particles in the atom that exists in nature.

Symbol	Formula	Value	Units
m_e			
The mass of an electron (Table 2)			
m_p			
The mass of a proton (Table 2)			
m_n	$m_e + m_p$	1.6735325768E-27	kg
The mass of a neutron [4]			
V_e	m_e / ρ_u	1.27822236702922E-47	m^3
The volume of an electron			
V_p	m_p / ρ_u	2.34700946985653E-44	m^3
The volume of an proton			
V_n	m_n / ρ_u	2.34828769222356E-44	m^3
The volume of an neutron			
r_e	$\sqrt{[\ 3.V_e\ /\ 4\pi\ ]}$	1.45046059426276E-16	m
The radius of an electron			
r_p	$\sqrt{[\ 3.V_p\ /\ 4\pi\ ]}$	1.77613270336827E-15	m
The radius of a proton			
r_n	$\sqrt{[\ 3.V_n\ /\ 4\pi\ ]}$	1.77645508248591E-15	m
The radius of a neutron			
t^N	$a_o\ /\ c$	1.765145168878310E-19	s
Newton time			
λ^N	a_o	5.291772106700000E-11	m
Newton length			
m^N	$a_o.c^2\ /\ G$	7.126607963504500E+16	kg
Newton mass (refer to Chapter 6.7)			
E^P	$m^N.c^2$	6.405075856756780E+33	J
Newton energy			
F^N	$E^P\ /\ \lambda^N$	1.210383918205250E+44	N
Newton force			
ρ_u	$\sqrt{[m_e.m_p]}\ /\ \sqrt{\Sigma}$	7.1266079635045E+16	kg/m^3
The ultimate density of Newton's atomic particles			
Table 6.10.1			

6.10.2 Planck's Atom (particle properties)

The following Table provides the properties of particles in a fictitious atom that has been constructed using Planck's values; t^P, λ^P, m^P

Symbol	Formula	Value	Units
m_e^P	m^P	2.1765500017459E-08	kg
Planck's electron mass, which is equal to Planck's mass			
m_p^P	m^P	2.1765500017459E-08	kg
Planck's proton mass, which is equal to Planck's mass (and Planck's electron)			
m_n^P	N/A	N/A	N/A
Planck's neutron is unnecessary as his atom is theoretical only			
V_e^P	$\sqrt{\Sigma}$	5.47722557505166E-46	m^3
Planck's electron volume			
V_p^P	$\sqrt{\Sigma}$	5.47722557505166E-46	m^3
Planck's proton volume			
V_n^P	N/A	N/A	N/A
Planck's neutron is unnecessary as his atom is theoretical only			
r_e^P	$\sqrt{[\,3.V_e^P / 4\pi\,]}$	5.07563837996471E-16	m
Planck's electron radius			
r_p^P	$\sqrt{[\,3.V_p^P / 4\pi\,]}$	5.07563837996471E-16	m
Planck's proton radius			
r_n^P	N/A	N/A	N/A
Planck's neutron is unnecessary as his atom is theoretical only			
t^P	$\sqrt{[\,\hbar.G / c^5\,]}$	5.39096122598359E-44	s
Planck time			
λ^P	$\sqrt{[\,\hbar.G / c^3\,]}$	1.61616952231128E-35	m
Planck length			
m^P	$\sqrt{[\,\hbar.c / G\,]}$	2.1765500017459E-08	kg
Planck mass (refer to Chapter 6.7)			
E^P	$m^P.c^2$	1.95618559889902E+09	J
Planck energy			
F^P	C^8 / G^2	1.21038391820525E+44	N
Planck force			
ρ_u^P	m^P / V_e^P	3.97381844498046E+37	kg/m^3
The ultimate density of Planck's atomic particles			
Table 6.10.2			

6.11   Important Constants (explained)

It is a fact that many natural constants are incorrect, approximate or misunderstood. It is the aim of this Chapter to finally resolve this issue.

Each of the following sub-Chapters describe the reasoning behind the units and values given for the most important physical constants you will find in the main body of the text.

6.11.1 Σ

This universal constant seems to tie everything together including Newton's and Planck's particles: Σ = 3E-91 (exact). Its importance is self-evident as it also gives us 'ρ_u' and 'G'. It has such a bizarrely accurate value, however, that it may or may not exist (see below). The trouble is, it seems to appear everywhere. E.g.:

$G = \sqrt{[\Sigma.a_o^2.c^4 / m_p.m_e]}$

$m_e.m_p = \Sigma.\rho_u^2$

$V_e.V_p = \Sigma$
$V^{P2} = r^{P6}.(^4/_3.\pi)^2 = \Sigma$

$r^P = {}^6\sqrt{[\Sigma / (^4/_3.\pi)^2]} = 5.075638379964711\text{E-}16$ m
is the radius of a Planck particle

If $F^N = G.m_e.m_p / a_o^2$ then; $F^N/F^P = \Sigma$
(refer to Chapter 6.10; Tables A to C; F^P)

Its units *appear* to be m^6. I say *"appears"* because it also unites the Newton and Planck forces (F^N & F^P; see above), which means it can also have no units.

Σ defines the ratio; "*electron orbital radius*" : "*particle centres*" (ϑ)
That is; the relative radii of the orbiting electron and the "electron radius plus proton radius" ($\vartheta = R / [r_e+r_p]$)
For example; if Σ = 3E-91, at a velocity of 'c'; ϑ = 1.46677550700175
The following Table shows the changes that would result in modifying Σ to give us ϑ = 1.0 (i.e. the particles touch at 'c').
All other physical constants unchanged.

Σ	3E-91	2.98746723133494E-90	{m⁶}
ϑ	1.46677550700175	1.0	
G	6.67359232004334E-11	2.10596243650527E-10	m³ / kg.s²
ρu	7.12660796350450E+16	2.25835348953929E+16	kg/m³
φ	4.407421117923340E-40	1.39083463164654E-39	
re	1.45046059426276E-16	2.12750007353581E-16	m
rp	1.77613270336827E-15	2.60518794648537E-15	m

Whilst; Σ = 3E-91 is not (yet) cast in stone, it does provide the most acceptable results for particle density, which is the basis for light-deflection and gives the correct value for G (refer to Chapter 6.11.2)

6.11.2 G

Today, you will see the units for this constant written as: $N.m^2/kg^2$, which were units of convenience originally assigned to reflect Newton's formula:
$F = G.m_1.m_2 / R^2$
This was because the formula for 'G' was unknown, hence its units were unknown.

From the relationship between Newton's Atom and Planck's Atom (refer to Chapter 6.11.1):
$F^N/F^P = \Sigma$ (no units)
where:
$F^N = G.m_p.m_e/a_o^2$ {N}
$F^P = c^4/G$ {N}
$G = \sqrt{[\ \Sigma.a_o^2.c^4 / m_p.m_e\ ]}$
$\quad = 6.67359232004334E\text{-}11$ {$\sqrt{[m^2.m^4 / s^4.kg^2]} = m^3 / s^2.kg$}
giving us its value and its units.

From Newton's laws of orbital motion, we know that a force-centre mass may be found thus: $m_1 = (2\pi)^2 / G.K$, and from Chapter 6.11.14 of this book, we know that: $K = (2\pi/v)^2 / R$
Therefore: $G = v^2.R/m_1$ confirming the units for 'G' as; $m^3 / kg.s^2$

If, on the other hand, Σ has units; m^6
$V_p = m_p/\rho_u$
$V_e = m_e/\rho_u$
$\Sigma = V_p.V_e = m_p.m_e/\rho_u^2$
$G.m_p.m_e/a_o^2 \div c^4/G = m_p.m_e/\rho_u^2$
$G^2 / a_o^2.c^4 = 1/\rho_u^2$
then;
$G = a_o.c^2/\rho_u$ {$m^6/kg/s^2$}
Because we know its value and units, 'G' must be a coefficient per unit volume $(m = V.\rho)$ giving us:
i.e. $G = a_o.c^2 / m_u$ {$m^3/kg/s^2$} per m^3
$G = a_o.c^2 / m_u$
$G = 5.2917721067E\text{-}11 \times 299792459^2 / 7.12660796350450E\text{+}16$
$\quad = 6.67359232004334E\text{-}11\ m^3 / s^2.kg$ per m^3

If, therefore, G = gravitational acceleration multiplied by spherical area per unit mass, the following calculations show that, contrary to popular belief, gravitational force does not vary with the square of the distance from the centre of its source.

Factors **1.5** & **4** below are exact values and will be explained in due course
*Note: A = **1.5**/a_o*

$G = \mathbf{1 \cdot 5}.c^2 / A.\rho_u = 6.67359232004334E-11$
$\{m^2/s^2 / (m^2.kg/m^3) = m^3 / s^2.kg\}$
$A = \mathbf{1 \cdot 5}.c^2 / G.\rho_u = 2.83458918818674E+10$
$\{m^2/s^2 / (m^3 / s^2.kg \cdot kg/m^3) = m^2\}$
$R = \sqrt{[A / 4.\pi]} = 47494.1512680647 \{m\}$ #
$V = {}^4/_3.\pi.R^3 = 4.48754692288540E+14 \{m^3\}$
$R_s = 2.G.m / c^2 = 47494.1512680647 \{m^3 / s^2.kg \cdot kg / (m^2/s^2) = m\}$ #
$R = R_s$
$m = R.c^2 / 2.G = 3.19809876372352E+31 \{m \cdot m^2/s^2 / (m^3 / s^2.kg) = kg\}$
$\rho = m/V = 7.12660796350450E+16 \{kg/m^3\}$
$\rho = \rho_u$

$G = R_s.c^2 / 2.m = 6.67359232004334E-11 \{m \cdot m^2/s^2 / kg = m^3 / s^2.kg\}$
$R_s = 2.G.m / c^2 = 47494.1512680647 \{m^3 / s^2.kg \cdot kg / (m^2/s^2) = m\}$ #
$g = G.m / R^2 = 9.46174592804013E+11 \{m^3 / s^2.kg \cdot kg / m^2 = m/s^2\}$
$F^N = G.m^2 / R^2 = 3.02595979551312E+43$
$\{m^3 / s^2.kg \cdot kg^2 / m^2 = kg.m/s^2 = N\}$
$F^P = c^4 / G = 1.21038391820525E+44$
$\{m^4/s^4 / (m^3 / s^2.kg) = kg.m/s^2 = N\}$
$F^P/F^N = \mathbf{4}$
$R_s = \sqrt[4]{[F^P / ({}^4/_3.\pi)^2.G.\rho^2.\mathbf{4}]} = 47494.1512680647$ #
$\{\sqrt[4]{[kg.m/s^2 / (m^3 / s^2.kg) / (kg^2/m^6)]} = m\}$
Schwarzschild radius (R_s) of mass 'm'

From the above formulas:
$G = g.A / 4.\pi.m = 6.67359232004334E-11 \{m^3 / s^2.kg\}$
$4.\pi.G = g.A / m = 8.38628344228057E-10 \{m^3 / s^2.kg\}$
Indicating that $F^N = 4.\pi.G.m_1.m_2 / 4.\pi.R^2$

In other words; Newton's gravitational force equation should read:
$F^N = G.m_1.m_2 / A$
and his gravitational constant should be:
$G = 8.38628344228057E-10 \ m^3/s^2/kg$;
i.e. gravitational force is constant irrespective of distance from the centre of its source, but it will vary at distance (R) according to its distribution over the spherical area, upholding the conservation of energy law.

Two aspects of this discovery require further explanation:
1) $a_o = \mathbf{1 \cdot 5}$ / A (A = 2.83458918818674E+10 m²); what does 'A' represent?
2) $F^P/F^N = \mathbf{4}$ (exact), which also applies to Newton's formula for the deflection of light (refer to Chapter 6.2.1) and surface area A = $\mathbf{4}.\Pi.R^2$ of a sphere over which Newton's and Coulomb's forces are spread.

Both of the above will be addressed in a later edition of this publication but neither of which alters the final result.

And finally: according to Planck: $G = 2\pi.\lambda^2.c^3$ / $h = \lambda^2.c^3$ / $\hbar$ {m³ / s².kg} confirming the units for 'G'.

In this solution for 'G', Planck's atom and Newton's theories have been fully analysed and complement each other perfectly. The discovery of so many atomic associations with G means that Newton actually did anticipate both Poincaré and Planck, confirming that his theories can be applied throughout all science, from the largest to the smallest.

i.e. there is no need for a unification theory (*Quantum theory is dead*)

6.11.3 φ

The coupling ratio is the ratio between gravitational (magnetic) charge (E_g) and electrical charge (E_e) and is defined as follows:
$E_g = G.m_1.m_2 / R$ and $E_e = k.q_1.q_2 / R$ {J}

$\varphi = E_g / E_e$
$\varphi = G.m_1.m_2 / k.q_1.q_2$

$$\varphi = \frac{(6.67359232004334E\text{-}11 \times 9.1093897E\text{-}31 \times 1.67262163783E\text{-}27)}{(8.98755184732667E\text{+}09 \times 1.60217648753E\text{-}19 \times 1.60217648753E\text{-}19)}$$

$\varphi = 4.40742111792334E\text{-}40$

This means that gravitational energy can only alter the density of matter if there is sufficient mass to generate the required pressure $(1/\varphi)$ and this only occurs in the ultimate-body

The following relationship is also true:

$\varphi = V_p.a_o / R_n$

V_p; Refer to Chapter 6.10.1
a_o; Refer to Chapter 5.4
R_n; Refer to Chapter 5.3

6.11.4 k, k', μ_o, ε_o

Coulomb's constant is equivalent to Isaac Newton's gravitational constant (G) when applied to *electrical* force in exactly the same fashion. I.e.:

Isaac Newton's formula for gravitational force:
$F = G.m_1.m_2 / R^2$

Charles-Augustin de Coulomb's formula for electrical force:
$F = k.q_1.q_2 / R^2$

The difference between the two is defined as the coupling ratio:
$\varphi = G.m_p.m_e / k.q_p.q_e = 4.40742111792335E-40$

Given the following ...
... for the magnetic constant:
$\mu_o = 4\pi.\mu'$ H/m
$\mu' = 1.0E-7 = \mu_o / 4\pi$ {H/m}
$\mu' = 1.0E-7 = m_e.R_n / e^2$ {kg.m/C^2}
$\mu_o = 4\pi.m_e.R_n / e^2$ {kg.m/C^2}
Therefore, the unit 'Henry' is actually; kg.m^2/C^2

... and permittivity in a vacuum:
$\varepsilon_o = 1 / \mu_o.c^2$ {C^2.s^2 / kg.m^3}

Coulomb's constant (k) may be calculated as follows:
$k = \mu_o.c^2 / 4\pi = 8.98755184732667E+09$
or
$k = \mu'.c^2 = m_e.R_n.(c/e)^2 = 1 / 4\pi.\varepsilon_o = \mu_o.c^2 / 4\pi$
$k = 8.98755184732667E+09$ {kg.m^3 / C^2.s^2}
When modified for electro-magnetic emission via protons:
$k' = k/\xi_m^2$

It is also interesting to note that the fine-structure constant is equal to:
$\alpha = e^2 / 4\pi = 2.h'.\varepsilon_o = 2.0427294212227E-39$ {C^2}

And the modified version of Planck's constant (refer to Chapter 6.11.5; h')
is related to Coulomb's constant thus:
$h' = e^2 / 8\pi.\varepsilon_o = \frac{1}{2}.k.e^2 = 1.15353857232684E-28$ {kg.m^3/s^2}

Moreover; electrical resistance x electron velocity is a constant:
311971413.341244 J.m/C^2 (irrespective of temperature)
Whereas Coulomb's constant k = 8.98755184732667E+09 J.m/C^2

6.11.5 h, h'

Because we know that energy cannot be created, it can only be transferred, the electro-magnetic energy emitted by a proton-electron pair must be the same as the kinetic energy in the orbiting electron. Moreover, the frequency (f) and amplitude (A) of electro-magnetic radiation is also equal to that of the orbiting electron (A = R and f = v / 2πR).

Max Planck claimed that the energy of electro-magnetic radiation can be calculated using his constant as follows:
$\mathbf{E^P}$ = h.f
in which his constant (h) is defined thus:
h = $\sqrt{[\pi.m_e.e^2.a_o / \varepsilon_o]}$ = 6.62607174469163E-34 **J.s**
However, these units (J.s) can only be considered correct if a frequency ratio is applied to his constant thus:
h = $\sqrt{[\pi.m_e.e^2.a_o / \varepsilon_o]}$. f_1/f_2
In reality, the units for Planck's constant should be:
$\sqrt{[kg.C^2.m / (C^2/m/(kg.m^2/s^2))]}$ = $\sqrt{[kg^2.m^4/s^2]}$ = **kg.m²/s**

h = $\sqrt{[\pi.m_e.e^2.a_o.4\pi.m_e.R_n.c^2 / e^2]}$
  = $\sqrt{[4\pi^2.m_e^2.c^2.a_o.R_n]}$
h = $\sqrt{[(4\pi)^2.a_o . R_n]}$. ½m_e.c {identical units; kg.m²/s}

Planck therefore *actually* identified a range of orbital radii:
a maximum: $(4\pi)^2.a_o$
a minimum: R_n
and a mean: $\sqrt{[(4\pi)^2.a_o . R_n]}$
The maximum orbital velocity: c

Because we know Planck's mean orbital radius: R_m = $2\pi.\sqrt{[(4\pi)^2.a_o . R_n]}$
R_m = 4.852618433622630E-12 m
and we also know the mean velocity: v_m = $\sqrt{[X_R / X.R_m]}$
v_m = 7224342.80705001 m/s
and we can calculate his minimum electron velocity (v_o) from:
v_o = c . $\sqrt{[R_n / (4\pi)^2.a_o]}$ = 174090.866621082 m/s

However, if we modify Planck's constant thus:
h' = h.v_o = 1.15353857232684E-28 **kg.m²/s . m/s**
his modified formula becomes:
h' = $\sqrt{[(4\pi)^2.a_o . R_n]}$. ½.m_e . v_o.c = ½.$R_n.m_e.c^2$ {kg.m³ / s² = J.m}
and because we can also calculate temperature (refer to Chapter 6.6), we can find everything we need to know about his orbiting electron:

Energy:	min	mean	max
R (m)	8.35643156381572E-09	4.85261843362268E-12	2.817937953839E-15 #
v (m/s)	174090.866621084	7224342.80705001	299792459
KE (J)	1.38042005551962E-20	2.37714666443634E-17	4.09355561131261E-14
PE (J)	-2.76084011103925E-20	-4.75429332887267E-17	-8.18711122262522E-14
T (K)	210.193328535837	361962.554671561	623316124.717179

Table 6.11.5-1: Planck's Electron Performance Range

Note: # the neutronic radius (R_n) is 2.81793795383896E-15 m

Whilst Max Planck's original formula for electro-magnetic energy (E^P = h.f) is incorrect, an alternative calculation method (E = h'/A) using the modified version of his constant (h') does work (refer to Table 6.11.5-2 below).

Refer to Chapter 4.1.3 for the properties of electro-magnetic radiation associated with the above temperatures.

Another interesting relationship with h':
e.k/h' = 2   {C² . kg.m³ / C².s² . s² / kg.m³ = no units}

Moreover; if we extract $\sqrt{[(4\pi)^2.a_o . R_n]}$ from **h** and modify it slightly, thus:
$\sqrt{[(4\pi)^2.a_o / R_n]}$ = 1722.0458764934
we get the velocity ratio ξ_v (refer to Chapter 5.3)

So, whilst Max Planck's claim regarding electro-magnetic energy using his constant was erroneous, without his work and proposal(s) these solutions would have been much more difficult to achieve; thank you Max!
h' = h.v_o = ½.k.e² = e² / 8.π.ε₀ = **½.R_n.m_e.c²** {kg.m³/s² }

In fact, Max Planck is the *only* 20th Century scientist without whom none of these scientific discoveries would have been possible.

T	R (A)	v	KE	E^P (error)	E (error)
6.2332E+08	2.8179E-15	299792459	*4.0936E-14*	1.121929E-11 (274.1)	4.09356E-14 (1)
3.1166E+08	5.6359E-15	211985280.7	*2.0468E-14*	3.966620E-12 (193.8)	2.04678E-14 (1)
2.0777E+08	8.4538E-15	173085256.9	*1.3645E-14*	2.159154E-12 (158.2)	1.36452E-14 (1)
1.5583E+08	1.1272E-14	149896229.5	*1.0234E-14*	1.402412E-12 (137.0)	1.02339E-14 (1)
1.2466E+08	1.4090E-14	134071263.5	*8.1871E-15*	1.003484E-12 (122.6)	8.18711E-15 (1)
1.0389E+08	1.6908E-14	122389758.9	*6.8226E-15*	7.633763E-13 (111.9)	6.82259E-15 (1)
8.9045E+07	1.9726E-14	113310898.8	*5.8479E-15*	6.057850E-13 (103.6)	5.84794E-15 (1)
7.7915E+07	2.2544E-14	·105992640.4	*5.1170E-15*	4.958275E-13 (96.90)	5.11694E-15 (1)
6.9257E+07	2.5361E-14	99930819.7	*4.5484E-15*	4.155294E-13 (91.36)	4.54840E-15 (1)
6.2332E+07	2.8179E-14	94802699.6	*4.0936E-15*	3.547852E-13 (86.67)	4.09356E-15 (1)
5.6665E+07	3.0997E-14	90390827.4	*3.7214E-15*	3.075222E-13 (82.64)	3.72141E-15 (1)
5.1943E+07	3.3815E-14	86542628.5	*3.4113E-15*	2.698943E-13 (79.12)	3.41130E-15 (1)
4.7947E+07	3.6633E-14	83147467.9	*3.1489E-15*	2.393594E-13 (76.01)	3.14889E-15 (1)
4.4523E+07	3.9451E-14	80122904.9	*2.9240E-15*	2.141773E-13 (73.25)	2.92397E-15 (1)
4.1554E+07	4.2269E-14	77406080.1	*2.7290E-15*	1.931206E-13 (70.77)	2.72904E-15 (1)
3.8957E+07	4.5087E-14	74948114.7	*2.5585E-15*	1.753015E-13 (68.52)	2.55847E-15 (1)
3.6666E+07	4.7905E-14	72710351.4	*2.4080E-15*	1.600635E-13 (66.47)	2.40797E-15 (1)
3.4629E+07	5.0723E-14	70661760.2	*2.2742E-15*	1.469118E-13 (64.60)	2.27420E-15 (1)
3.2806E+07	5.3541E-14	68777107	*2.1545E-15*	1.354675E-13 (62.88)	2.15450E-15 (1)
3.1166E+07	5.6359E-14	67035631.7	*2.0468E-15*	1.254355E-13 (61.28)	2.04678E-15 (1)
2.9682E+07	5.9177E-14	65420077.9	*1.9493E-15*	1.165834E-13 (59.81)	1.94931E-15 (1)
2.8333E+07	6.1995E-14	63915967	*1.8607E-15*	1.087255E-13 (58.43)	1.86071E-15 (1)
2.7101E+07	6.4813E-14	62511048.9	*1.7798E-15*	1.017124E-13 (57.15)	1.77981E-15 (1)
2.5972E+07	6.7631E-14	61194879.4	*1.7057E-15*	9.542204E-14 (55.94)	1.70565E-15 (1)
2.4933E+07	7.0448E-14	59958491.8	*1.6374E-15*	8.975436E-14 (54.81)	1.63742E-15 (1)
2.3974E+07	7.3266E-14	58794138.4	*1.5744E-15*	8.462633E-14 (53.75)	1.57444E-15 (1)
2.3086E+07	7.6084E-14	57695085.6	*1.5161E-15*	7.996867E-14 (52.75)	1.51613E-15 (1)
2.2261E+07	7.8902E-14	56655449.4	*1.4620E-15*	7.572312E-14 (51.79)	1.46198E-15 (1)
2.1494E+07	8.1720E-14	55670062.1	*1.4116E-15*	7.184037E-14 (50.89)	1.41157E-15 (1)
2.0777E+07	8.4538E-14	54734364.1	*1.3645E-15*	6.827845E-14 (50.04)	1.36452E-15 (1)
2.0107E+07	8.7356E-14	53844315.1	*1.3205E-15*	6.500144E-14 (49.22)	1.32050E-15 (1)
1.9479E+07	9.0174E-14	52996320.2	*1.2792E-15*	6.197843E-14 (48.45)	1.27924E-15 (1)
1.8888E+07	9.2992E-14	52187168.5	*1.2405E-15*	5.918268E-14 (47.71)	1.24047E-15 (1)
1.8333E+07	9.5810E-14	51413982.6	*1.2040E-15*	5.659097E-14 (47.00)	1.20399E-15 (1)
1.7809E+07	9.8628E-14	50674174.5	*1.1696E-15*	5.418305E-14 (46.33)	1.16959E-15 (1)
1.7314E+07	1.0145E-13	49965409.8	*1.1371E-15*	5.194118E-14 (45.68)	1.13710E-15 (1)
1.6846E+07	1.0426E-13	49285576.7	*1.1064E-15*	4.984975E-14 (45.06)	1.10637E-15 (1)
1.6403E+07	1.0708E-13	48632758.7	*1.0773E-15*	4.789500E-14 (44.46)	1.07725E-15 (1)
1.5982E+07	1.0990E-13	48005213	*1.0496E-15*	4.606474E-14 (43.89)	1.04963E-15 (1)
1.5583E+07	1.1272E-13	47401349.8	*1.0234E-15*	4.434816E-14 (43.33)	1.02339E-15 (1)
1.5203E+07	1.1554E-13	46819716.1	*9.9843E-16*	4.273560E-14 (42.80)	9.98428E-16 (1)
1.4841E+07	1.1835E-13	46258980.7	*9.7466E-16*	4.121845E-14 (42.29)	9.74656E-16 (1)
1.4496E+07	1.2117E-13	45717921.4	*9.5199E-16*	3.978899E-14 (41.80)	9.51990E-16 (1)
1.4166E+07	1.2399E-13	45195413.7	*9.3035E-16*	3.844028E-14 (41.32)	9.30354E-16 (1)
1.3851E+07	1.2681E-13	44690421.2	*9.0968E-16*	3.716608E-14 (40.86)	9.09679E-16 (1)
1.3550E+07	1.2963E-13	44201986.6	*8.8990E-16*	3.596076E-14 (40.41)	8.89903E-16 (1)

Table 6.11.5-2: Planck's Energy (E^P = h.f)

Note: the above (error) is a ratio and therefore a value of 1 represents zero error

6.11.6 e, e'

The electrical charge in an electron is 1.60217648753E-19 C and designated the symbol 'e'. This charge is invariable; it never varies in an electron. This is not quite the same for the proton, however;

The charge generated in a proton by its orbiting electron is a constant and may be defined as follows:

e' = e.ξ_v.$\sqrt{[T/T_n]}$ = 2.75902141376572E-16 C

The static charge in a lone proton is always the same as that in an electron (e), but contrary to an electron, when a proton attracts an orbiting electron, its own charge will increase to e'. This is facilitated by the proton's surplus non-polar magnetic capacity (mass). I.e.:

The electrical charge in a lone proton (and in an electron) is always; e

The electrical charge in a proton with an orbiting electron rises to; e'

e' is also the maximum electrical charge in all the electro-magnetic radiation emitted by the proton

6.11.7 R_∞, R_γ, a_o

What is commonly regarded as Bohr's radius is the orbital radius of an electron that possesses the kinetic energy equal to Johannes Rydberg's predictions for an electron at what was then called an electron's ground state. 'a_o' is not, however, the ground state orbital radius of an electron.

$a_o = 4\pi.\varepsilon_o.h^2 / m_e.e^2 = (h / 2\pi.m_e.c)^2 / R_n$ {m}

If the ground state of an electron can reasonably be said to occur at the point it ceases to provide sufficient energy to hold on to a neutron, this will be when the electron is orbiting:
at a radius of 8.3564315638157900E-09 m
at a temperature of 210.193328535837 K

Rydberg generated the following formula for his first constant:
$R_\infty = m_e.e^4 / 8.\varepsilon_o^2.h^3.c = 10973726.9561356$ {/m}
which breaks down to:
$R_\infty = \sqrt{R_n} / 4\pi.a_o^{1.5} = 10973726.9561356$ {/m}
and is his wave number (for electro-magnetic energy).

All of which breaks down to; $R_\infty = 1 / a_o.\xi_v$ {/m}

Rydberg generated the following formula for universal energy constant for an electron:
$R_\gamma = R_\infty.h.c.(Z.n)^2$
which breaks down to:
½ . R_n/a_o . $m_e.c^2 = 2.17987197684933E-18$ {J}
and relates to a temperature of 33192.4000063507 K
However, this value is of no significance to a key natural event.

But Planck's minimum orbital radius in his own constant 'h' is; $[4\pi]^2.a_o$
(refer to Chapters 5.4 & 6.11.5)

And if Rydberg had modified his formula to reflect Planck's value:
$R_\gamma = ½ . (R_n / [4\pi]^2.a_o) . m_e.c^2 = 1.38042005551962E-20$ {J}
Rydberg would have revealed the electron's minimum Planck energy 40 years earlier. Refer to Chapter 3.5.4

It is not yet understood what relevance Rydberg's radius (a_o) has to the atomic structure, because there is no such thing as 'rest-mass' for electrons as they never come to rest.

6.11.8 ρ_u

The ultimate (limiting) density is the maximum possible density achievable in nature, and it applies to both atomic particles; the electron and the proton.

ρ_u = 7.12660796350449E+16 kg/m³

It may be calculated as follows:

$\rho_u = \sqrt{[m_e.m_p]} / V_e^P$
Refer to Chapter 6.10.2 for V_e^P

or

$\rho_u = a_o.c^2 / G$
Refer to Chapter 6.11.2

or

$\rho_u = \sqrt{[m_e.m_p / \Sigma]}$

or

$\rho_u = m_e / \sqrt{[\Sigma . m_e/m_p]}$

or

$\rho_u = m_p / \sqrt{[\Sigma . m_p/m_e]}$

6.11.9 R_s

Schwarzschild's radius is the radius of a spherical body, the non-polar magnetic (gravitational) energy of which, is sufficient at its surface to trap the mass of an electron travelling at light-speed.

It is calculated as follows:

$R_s = 2.G.m/c^2$
Refer to Chapter 5 for definitions of 'G' and 'c'
m is the mass of the spherical body

The following examples give some idea of the Schwarzschild radii of various objects:

$1m^3$ of matter of ultimate density (7.12660796350450E+16 kg):
$R_s = 2.G.m_p/c^2 = 1.05835442134E-10$ m
i.e. you don't need very much of this matter to trap an electron travelling at light-speed.

The minimum sized black-body of iron density (ρ_i):
$R_s = c.\sqrt{[\ 3\ /\ 8.\pi.G.\rho_i\ ]} = 1.42875013455622E+11$ m
$m = {}^4/_3\pi R_s{}^3 . \rho_i = 9.6237854E+37$ kg

A proton:
$R_s = 2.G.m/c^2 = 2.48396784934951E-54$ m
Proving that a proton cannot trap an electron through gravitational force alone. But if the coupling ratio were applied, obtaining potential energy through electrical charge;
$R_s = 2.G.m\ /\ \varphi.c^2 = 5.63587590767792E-15$ m
and the neutronic radius is: $R_n = 2.81793795383896E-15$ (exactly half!)

Given that the Schwarzschild's radius is sized to trap kinetic energy, and potential energy is exactly twice kinetic energy in orbiting electrons, half Schwarzschild's radius is the limiting potential electrical [charge] energy required to trap an electron travelling at light-speed.

A Schwarzschild's radius is, however, an hypothetical value as an electron travelling at velocity 'c' is supposed to represent a photon, which doesn't exist, so trapping a *photon* is an unreal scenario.

6.11.10 R_n

The neutronic radius is the orbital radius of an electron when it is travelling at the speed of electro-magnetic radiation (light-speed). At this speed, the electron and proton combine to become a neutron.

$R_n = \mu'.e.RC$ { kg.m / C^2 . C . C/kg = m }

According to Newton's orbital motion formula; $v = \sqrt{[R.g]}$
'v' will be the speed of light (c) (Table 10)
when:
R_n = orbital radius of 1.46677550700177 x $(r_p + r_e)$
 = 2.817937953839E-15 m
$g = G.m_p / \varphi.R_n^2 = 3.18940728807829E+31$ m/s^2

Moreover, if:
$T = X.v^2 = X_R/R$
$R = X_R / X.v^2$
At light-speed (c)
$R_n = X_R / X.c^2 = 2.81793795383896E-15$ m

According to the relationship: $T = X.v^2$
at a temperature of $T = X.c^2 = 623316124.71718$ K
The reason an electron is trapped at this radius, is due to it venturing inside the Schwarzschild radius of the proton at this velocity (refer to Chapter 6.11.9).

6.11.11   RAC & RAM

Relative atomic charge and relative atomic mass define the capacity of a particle to hold an electrical charge (e) or *mass* (m) respectively.
They are related as follows:

RAM {g/mol}

RAC {C/mol}

If either of the above is divided by Avogadro's constant (N_A), you will get the capacity (m, e) of the particle {g, C}

If ideal gas constant (R_i) is divided by either of the above, you will get the *specific* capacity (c, q) of the particle {J/g/K, J/C/K}

If the capacity is multiplied by the *specific* capacity, you will get the *relative* capacity (C, Q) of the particle {J/K}

6.11.12 N_A

Avogadro's number is the number of C^{12} (pure carbon) atoms in 12g and is recognised as; N_A = 6.02214129E+23 /mole

However, one atom of pure carbon-12 has a mass of:
m_C = 6.$(m_e+m_p+m_n)$ = 2.00823909216000E-23 g

i.e.; N_A = 1/(m_e+m_p) {/mol}
where m_e & m_p are specified in grams

and 12 grams of pure carbon-12 contains:
N_A = 12/m_C = 5.97538412973187E+23
which is 0.7764208% less than Avogadro's number

If corrected, this would, of course alter a number of constants such as:
R_i; X; X_R; c & q

However, because this 0.78% inconsistency has little effect on the Newton's laws of motion, the corrected figure has not been adopted here. Avogadro's Number has been left as he defined it.

6.11.13 k_B

Boltzmann's constant describes the potential energy per unit temperature between a proton and its orbiting electron, and comprises the following fundamental constants:

$k_B = m_e.c^2 / Y.T_n = 1.38065156E\text{-}23 \text{ J/K}$

Together with Avogadro's constant, Boltzmann's constant defines the ideal gas constant:

$R_i = N_A.k_B = 8.24992342031355 \text{ J/K/mol}$

Notes:
$e^2/4\pi$ is what is commonly referred to as the *'fine structure constant'* (α)

An interesting solution for Boltzmann's constant is described below:
$\frac{1}{3} . m_n . v_C^2 = \mathbf{1.3806}4741873833E\text{-}23 \{J/K\}$
$k_B = \mathbf{1.3806}5156E\text{-}23 \{J/K\}$
Where:
$\frac{1}{3} = 1 / 3K$ {temperature}
m_n = mass of a neutron
v_C = velocity constant (refer to Chapter 6.11.15)

6.11.14   K

This constant of proportionality is common to all orbits.

In elliptical orbits, its value is governed by the *force-center's* mass and its formula is not only;

$K = t^2/a^3 = (2.\pi)^2 / G.m_1$

but also;

$K = 2\pi/v_{max} \cdot 2\pi/v_{min} / a \ \{s^2/m^3\}$

Where:
v_{max} = maximum orbital velocity
v_{min} = minimum orbital velocity
a = half the orbital major axis

In circular orbits, its value is governed by the *satellite's* mass, and its formula alters to:

$K = (2\pi/v)^2 / R \ \{s^2/m^3\}$
Where:
v = orbital velocity
R = orbital radius

Its value can be any number greater than 0

For a proton: K = 0.15587874533403

For our lunar system: K = 9.91826542816423E-14

For our solar system: K = 2.97491436434708E-19

For our Milky Way: K = 3.35025744566253E-30
Assuming an orbital eccentricity of 0.015941744 for our solar system

6.11.15 v_C & v_E

Two velocity constants (v_C & v_E) have been identified, but neither have yet been fully resolved in terms of their orbital applicability.

If $E = 3.k_B/m_n$:
$v_C = \sqrt{[\ k_B.3/m_n\ ]} = 157.320597065167$ {m/s}
or
$v_C = \sqrt{[\ g.R/PE\ .\ (E_1.E_2)/(E_1+E_2)\ ]} = \mathbf{157.320361123578}$ {m/s}
where:
$E_1 = E.m_{sun}$
$E_2 = E.m_{earth}$;
m_n = mass of a neutron
m_{sun} = mass of our sun
m_{earth} = mass of the earth

Using the values for a proton-electron pair and its neutron as its mass is most accurately known:
$v_E = \sqrt{[\ k_B'.T/m_e\ ]}$
@ 1K: $v_E = \mathbf{12007.8850946582}$ {m/s}
Where: k_B' is the charge level of a proton (refer to Chapters 5.4 & 6.11.13)

The velocity of any electron in any shell is found by applying the correct temperature value (T) in the above formula.

Whilst both of the above velocities are genuine constants that can be applied to all orbits (circular and elliptical), their resolution will feature in a later revision of this book.

6.12 The Laws of Thermodynamics

The First Law of Thermodynamics: Conservation of energy
Energy can never be lost, it can only be transformed or transferred.

The Second Law of Thermodynamics: Heat will not spontaneously pass from a colder body to hotter body

A high-energy source (hotter body) will spontaneously lose energy to a low-energy source (colder body) but you must add work if you want energy to transfer in the other direction (up-hill so to speak). This law essentially states that it is impossible to create energy from nothing.

This law also claims that energy can, and in fact is, lost by a system to its surroundings but that the reverse cannot happen i.e. an increase in disorder is an inevitable feature of time.

The Third Law of Thermodynamics: The entropy of a substance approaches zero as its temperature approaches zero (absolute)

Entropy is the term used to define disorder. The higher a substance's temperature the more disordered will be its atomic structure and the higher its entropy. E.g. gas has a higher entropy than a solid substance.

7 Things You Can Do

After two and a half years intensive study of this subject I'm knackered! I have left the last bits for somebody else to solve: you perhaps!

[1] Neutrons

Refer to: *The Life & Times of a Neutron*; Keith Dixon-Roche; published by CalQlata

[2] Nucleic Structure

The nucleic structure (ζ : refer to Chapter 3.5.3) is mirrored by the lattice structure of an atom in both viscous and its gaseous forms. The mathematical association of ζ & Γ with the lattice are yet to be resolved.

[3] T_o

The minimum Planck temperature may be related to neutron retention. This needs to be confirmed by experiment.

[4] T_c

The temperature at which a proton is anticipated to lose its orbiting electron. This needs to be confirmed by experiment.

[5] Electrons in Free-Flight

Whilst electrons cannot *lose* kinetic energy whilst in free-flight, it is not known whether they can collect energy from surrounding radiation. This needs to be confirmed by experiment.

[6] Electro-Magnetic Radiation

It is currently *assumed* that an electro-magnetic wave can donate *part* of its energy thereby altering its characteristics. This needs to be confirmed.

[7] Negative Velocity

It is possible, but by no means probable, that the negative-positive balance between all properties means that a negative velocity must also exist. This would of course mean that it could be possible to travel backwards; i.e. reverse time! But it needs to be proven.

Appendices

References, symbols, glossary, etc. used throughout this book along with a summary list of corollaries and hypotheses.

A1 General

N/A

A2 References

Most of the references used for the creation of this book are from original work supplied in CalQlata (www.calqlata.com), but some additional sources are listed below:

Magnificent Principia; Colin Pask; 978-1-61614-745-7

Seven Brief Lessons on Physics; Carlo Rovelli; 978-0-141-98172-7

Science Data Book; Open University; 0 05 002487 6

Science and Technology Dictionary; Chambers; 0-550-18026-5

A Dictionary of Scientific Units; H G Jerrard & D B McNeill; 0-412-28100-7

It is important to note here that most of the sources here are from work done by pre-20th Century scientists that are universally known and available from sources too numerous to mention here.

A3 Glossary

Atomic Particle	One of the three components that comprise an atom
Big-Bang	The eruption that occurred when the Ultimate-Body accumulated sufficient 'mass' to compromise the integrity of the innermost neutron.
Black-Body	A collection of Quanta too cold to emit electro-magnetic radiation in the frequencies that would enable detection.
Coupling Ratio (φ)	The ratio of the coupling force due to a magnetic charge and the coupling force due to an electric charge: $\varphi = G.m_p.m_e \div k.e^2$ (refer to Chapter 6.11.3)
Gas	Atoms that possess greater electrical field energy than magnetic field energy
Hades	The Milky Way's force-centre
Proton-electron pair	A proton that hosts an orbiting electron
Sub-Atomic Particle	The many particles said to compromise atomic particles (leptons, gluons, fermions, quarks, etc.)
Ultimate-Body	A body that contains all the Quanta in the universe ($\approx 2.8E+75$) and represents the maximum single 'mass' that can exist without generating a Big-Bang.
Ultimate Density	The mass-density of all three atomic particles $\rho = 7.12660796350449E+16 \ kg/m^3$ Nothing in nature has a 'mass'-density greater than this value
Universal Period	The time elapsed since the last Big-Bang or between subsequent Big-Bangs
Viscous	Solid or liquid matter in which magnetic field energy is greater than an electron's electrical charge

All other definitions can be found on the following web page:
http://calqlata.com/help_definitions.html

A4 Symbols

Refer to Chapter 5 for a list of all the symbols used in this book.

The most prominent subscripts are listed below:

mass	e	electron
	p	proton
temperature	c	cold
	o	minimum Planck
	m	mean Planck
	n	maximum Planck
Rydberg	y	energy constant
	∞	wave number
	o	orbital radius (a_o) *occasionally referred to as the Bohr radius*
Others	u	Ultimate
radii	n	Neutron orbit radius
	s	Schwarzschild radius
	1	force-centre
	2	satellite
energy	e	electron
	p	proton

The most prominent superscripts are listed below:

Force	N	Newton
	P	Planck

A5 Useful Formulas

Equidistant arc-length between 'n' points on the surface of a sphere:
$d = \pi.A / C.n$
where C is the circumference of the sphere
Linear distance across arc-length 'd' (above):
$\ell = 2.R.Sin(\frac{1}{2}.d/R)$
but if you know 'ℓ' and need to find 'n':
$n = \pi / Asin(\frac{1}{2}.\ell/R)$
and if $\ell = R$:
$n = \pi / Asin(\frac{1}{2}) = \textbf{6}$

Lorentz's Equation (magnetic force or field strength):
$F = q.v.B$
Which becomes:
$F = q.g.R.B$
for the laws of orbital motion
Where:
q is the total electrical charge = $q_1.q_2 / m_e.(q_1+q_2)$
v = relative velocity (electrical circuits)
g = gravitational attraction between m_1 & m_2
R = radial separation between m_1 & m_2
$B = \mu_o.I / 2.\pi.R$ kg/C {$R = 2.R_n$}
$I = e$
$B = \mu_o.e / 4.\pi.R_n = 4.\pi.R_n.m_e/e^2 . e / 4.\pi.R_n = m_e/e = 1/RC$ kg/C
RC_e is the relative atomic charge of an electron {C/kg}
$B = 1/RC = 5.685634367312E-12$ kg/C

A6 Corollaries

Corollary 1: Everything in the universe is composed of electrical and magnetic energy

Corollary 2: Magnetic energy is accrued and travels from positive to negative

Corollary 3: Electrical energy is shared and travels from negative to positive

Corollary 4: Atoms comprise collections of proton-electron pairs

Corollary 5: A neutron is a proton-electron pair united under high temperature and holds 4.0935556113127E-14 J of energy

Corollary 6: Atoms exist in solid/liquid state (attraction) due to the magnetic field generated by its proton-electron pairs

Corollary 7: Atoms exist in gaseous state (repulsion) due to the electrical field generated by its proton-electron pairs

Corollary 8: All matter is either viscous (solid/liquid) or gaseous, dependent upon the dominance of its magnetic or electrical fields

Corollary 9: Mass is non-polar magnetic charge

Corollary 10: Gravity is the attractive force between magnetic charges

Corollary 11: Light is electro-magnetic energy

Corollary 12: Every orbital system *must* have a force-centre

Corollary 13: Isaac Newton's gravitational constant may be defined as follows:
$G = a_o.c^2/\rho_u$ {m^3 / s^2.kg per m^3}
Where: ρ_u is the ultimate density (7.12661E+16 kg/m^3)

Corollary 14: Potential energy remains constant irrespective of distance from its source

Corollary 15: Orbital shape is defined *only* by force-centre mass

Corollary 16: Kinetic energy of a satellite in Newton's laws of orbital motion is exclusive of that generated by its angular velocity

Corollary 17a: $PE/KE = -2.(1-e)/(1-e^2)$ for all orbits
Corollary 17b: The potential energy between a satellite and its force-centre is always twice the kinetic energy in the satellite for circular orbits

Corollary 18: The internal pressure of any mass can be calculated using Isaac Newton's force-formula thus:
$p = G.m_1.m_2 / A^2$, in which 'A' is the spherical area at radius 'r' from its centre (G must be in its modified form: 8.3862834423E-10 m^3 / s^2.kg).

Corollary 19: The centrifugal force on an orbiting body:
@ the perigee of an ellipse; $F_c = F / (1+e)$
@ the apogee of an ellipse; $F_c = F . (1+e)$

Corollary 20: The centrifugal velocity (v_c) of a satellite at any point in an elliptical orbit is:
$v_c = \zeta.v$
Where:
$\zeta = \sqrt{[(f.Sin(\theta/2)^a + p.Cos(\theta/2)^a) / (f.cos(\theta/2)^a + p.Sin(\theta/2)^a)]}$
$a = \sqrt{[^4/_3.\pi]}$
v = the satellite's elliptical velocity at the same point.

Corollary 21: A satellite's spin is defined by its force-centre's spin and its sub-satellite orbits

Corollary 22a[#]: Prograde spin is induced in a satellite by the potential energy between its centre and that of its force-centre
Corollary 22b[#]: Retrograde spin is induced in a satellite by prograde spin energy in its force-centre
Corollary 22c[#]: Prograde spin is induced in a force-centre by the sum of the perigee kinetic energies plus apogee potential energies of its satellite(s)
Assuming a satellite's orbit is in the prograde direction[#]:

Corollary 23: The difference between the spin rates in a satellite's core and its mantle is due to the conflicting influences of corollary 22

Corollary 24a: A satellite's internal heat is generated by the friction due to corollary 23
Corollary 24b: Heat can be generated within a satellite orbiting a force-centre with a period different to the force-centre's period of spin
Corollary 24c: Heat is increased within a satellite that has satellites of its own
Corollary 24d: A gas-planet is a satellite with sufficient internal heat to prevent a surface crust forming

Corollary 25: A satellite's mantle plumes are generated by its internal frictional heat.

Corollary 26: A satellite's magnetic field is generated by the differential spin rates in its core and its mantle.

Corollary 27: The angular difference between true and magnetic axes in a satellite with iron core and mantle that rotate at different rates can be calculated thus:

$a = sign(|\omega/\omega_m|) \cdot ½.\sqrt{[\ Asin(|\omega/\omega_m|)\ ]}$

Corollary 28: All stars produce hydrogen gas in the form of proton-electron pairs, by converting proton-electron pairs (in its body-matter) to neutrons using the frictional heat generated by planetary spin

Corollary 29: Any satellite may be replaced in any orbit without altering the orbital shape and period (velocities)

Corollary 30: A black-body is a celestial body of any mass or size that is too cold to emit electro-magnetic radiation

Corollary 31: The force-centre at the heart of the Milky Way galaxy (Hades) has a mass of ≈1.8E+41 kg, a diameter of 3.5E+12 m and is spinning at ≈2E-07 ᶜ/s according to NASA data on the Milky Way

Corollary 32: The density of the matter beneath the earth's crust is little more than that of water

Corollary 33: The coupling ratio (φ), that of magnetic to electrical potential energy is; 4.407E-40

Corollary 34: There is insufficient pressure at the core of a minimum celestial black-body to alter the density of matter.

Corollary 35: Satellites in circular orbits generate their own kinetic energy.

Corollary 36a: All heat is radiated
Corollary 36b: Conduction is the transfer of electro-magnetic energy between electrons within matter irrespective of its state
Corollary 35c: Convection is the movement of gaseous atoms to balance electrical repulsive forces (between adjacent atoms) with gravitational forces

Corollary 37: $E=mc^2$ applies to potential energy in circular orbits

Corollary 38: Mass remains constant irrespective of velocity

Corollary 39: There is no such thing as a photon
Electrons in free-flight do not emit light – light possesses no mass

Corollary 40: There is no such thing as dark matter in the form of sub-atomic particles